IMMOBILIZED ENZYMES

Contributors

Ichiro CHIBATA
  Director, Research Laboratory of Applied Biochemistry, Tanabe
  Seiyaku Co., Ltd.
  16-89 Kashima 3-chome, Yodogawa-ku, Osaka 532, Japan

Tetsuya TOSA
  Senior Scientist, Research Laboratory of Applied Biochemistry,
  Tanabe Seiyaku Co., Ltd.
  16-89 Kashima 3-chome, Yodogawa-ku, Osaka 532, Japan

Tadashi SATO
  Department of Biochemistry, Research Laboratory of Applied
  Biochemistry, Tanabe Seiyaku Co., Ltd.
  16-89 Kashima 3-chome, Yodagawa-ku, Osaka 532, Japan

Takao MORI
  Department of Biochemistry, Research Laboratory of Applied
  Biochemistry, Tanabe Seiyaku Co., Ltd.
  16-89 Kashima 3-chome, Yodogawa-ku, Osaka 532, Japan

IMMOBILIZED ENZYMES

Research and Development

Edited by Ichiro CHIBATA
Director, Research Laboratory of Applied Biochemistry,
Tanabe Seiyaku Co., Ltd., Osaka

A HALSTED PRESS BOOK

KODANSHA LTD.
Tokyo

JOHN WILEY & SONS
New York–London–Sydney–Toronto

KODANSHA SCIENTIFIC BOOKS

Library of Congress Cataloging in Publication Data

Main entry under title:

Immobilized enzymes, research and development.

  (Kodansha scientific books)
  "A Halsted Press book."
  Includes bibliographies.
  1. Immobilized enzymes. 2. Immobilized
enzymes--Industrial applications. I. Chibata,
Ichirō.
QP601.I474   574.1'925   78-13266
ISBN 0-470-26531-0

Published in Japan by
KODANSHA LTD.
12–21 Otowa 2-chome, Bunkyo-ku, Tokyo 112, Japan

Published by
HALSTED PRESS
a Division of John Wiley & Sons, Inc.
605 Third Avenue, New York, N.Y. 10016, U.S.A.

PRINTED IN JAPAN

Contents

Contents

Preface

With the development of studies on enzyme structure and function, the highly specific nature of enzyme action has become increasingly apparent. However, enzymes are not always ideal catalysts for particular industrial or academic purposes. In order to develop better catalysts for various applications, attempts have been made recently to prepare catalysts having enzyme-like activities by using new techniques of synthetic chemistry and polymer chemistry. Some of these catalysts are called "synzymes". Immobilization of enzymes, as described in this book, is another approach to obtaining better catalysts by modifying enzymes produced by organisms to make them more suitable for the desired purposes.

The authors first began work on immobilized enzymes nearly 15 years ago, and in 1969 succeeded in industrializing a process for the continuous optical resolution of DL-amino acids using an immobilized enzyme preparation. In 1973 we also succeeded in the industrialization of a continuous process for the production of L-aspartic acid using immobilized microbial cells. These processes are thought to be the first industrial applications of immobilized enzyme and immobilized microbial cell systems, respectively, in the world. Since then, many reports on immobilized enzymes have been published, many symposia and international conferences have been held, and the number of investigators working on immobilized enzymes has increased greatly. This field is multidisciplinary, and not only enzyme chemists and biochemists but also chemical engineers have participated in studies on immobilized enzymes, as chemical engineering techniques can be rather readily applied to immobilized enzyme systems, unlike conventional fermentation or enzymic methods, which require special biological knowledge and techniques. At present, immobilized enzymes are important tools for biochemical engineering and enzyme technology, and their applications for both academic and industrial purposes are expected to give them an increasingly important role in the life sciences.

Many reviews and books have been published on immobilized enzymes, but most deal with specific aspects, and are not comprehensive. We therefore have aimed to produce a comprehensive text covering the principles and applications of immobilized enzymes and immobilized microbial cells, based on the editor's experience in research on, and industrialization of, immobilized enzyme and immobilized microbial cell systems. The

vii

science and technology of immobilized systems now appears to have completed the early developmental stage, and may well be at a turning point. It is hoped therefore that this book will be useful as a handbook for workers in many disciplines, both scientists and engineers, and that it will contribute to some degree to the further development of this field.

Finally, as editor for this project, I wish to express my appreciation of the work of the contributing authors. I would also like to take this opportunity to thank Mr. W.R.S. Steele and the staff of Kodansha for invaluable assistance in the preparation of the English manuscript.

March 1978

Ichiro CHIBATA
Osaka, Japan

CHAPTER 1

Introduction

1.1 BACKGROUND

Enzymes are biological catalysts, consisting of protein, which participate in many chemical reactions occurring in living things. Unlike ordinary chemical catalysts, enzymes characteristically have the ability to catalyze a reaction under very mild conditions in neutral aqueous solution at normal temperature and pressure, and with very high specificity.

Enzymes have been utilized by human beings since ancient times, well before their nature was understood. The use of enzymes has gradually been extended into a variety of fields, such as brewing, food production, textiles, tanning, and medicine. Further, the recent development of biochemistry has resulted in the clarification of mechanisms of enzyme reaction and the development of new enzyme sources, and, together with progress in applied microbiology, has greatly extended the range of application of enzymes. However, enzymes are produced by organisms for their own requirements, and although enzymes are efficient and effective catalysts, they are not always ideal for practical applications. Some of the advantages of enzymes may even be disadvantages in practical use as catalysts. Namely, enzymes are generally unstable, and cannot be used in organic solvents or at elevated temperature.

Conventionally, enzyme reactions have been carried out in batch processes by incubating a mixture of substrate and soluble enzyme. It is technically very difficult to recover active enzyme from the reaction mixture after the reaction for reuse. Accordingly, the enzyme and other contaminating proteins are generally removed by denaturation by pH or heat treatment during procedures to isolate the product from the reaction mixture. This is uneconomical, as active enzyme is lost after each batch reaction.

To eliminate the disadvantages inherent in ordinary chemical catalysts and enzymes, and in order to obtain superior catalysts for applications, that is, highly active and stable catalysts having appropriate specificity, two approaches have been investigated. One is the synthetic approach, using

1

recently developed techniques of organic synthesis and polymer chemistry to synthesize catalysts having enzyme-like activities. These catalysts are sometimes called "synzymes." Another approach is the modification of enzymes produced by organisms. Immobilization of enzymes is a part of the latter approach.

If active and stable water-insoluble enzymes, i.e., immobilized enzymes having appropriate substrate specificity, are prepared, most of the above disadvantages are eliminated and it becomes possible to use enzymes conveniently in the same way as ordinary solid catalysts used in synthetic chemical reactions. In addition, since enzymes can catalyze specific reactions under mild conditions, (normal temperature and pressure), application of immobilized enzymes in the synthetic chemical industry can reduce energy requirements.

Immobilized enzymes are considered to be modifications of the enzymes, and should also be useful in clarifying the relationship between protein structure and enzyme activity or reaction mechanism. Many enzymes are believed to be bound to cell membranes or cellular particles in organisms. Immobilized enzymes are considered to be models of these bound enzymes, and are becoming the subject of extensive academic interest.

In 1916, Nelson and Griffin[1] found that an enzyme in water-insoluble form showed catalytic activity. They reported that invertase extracted from yeast was adsorbed on charcoal, and the adsorbed enzyme showed the same activity as native enzyme. In 1948, Sumner[2] found that urease from jack bean became water-insoluble on standing in 30% alcohol and sodium chloride for 1–2 days at room temperature, and the water-insoluble urease was active. Thus, it has been known for a long time that enzymes in water-insoluble form show the catalytic activity. However, the first attempt to immobilize an enzyme to improve its properties for a particular application was not made until 1953, when Grubhofer and Schleith[3] immobilized enzymes such as carboxypeptidase, diastase, pepsin and ribonuclease by using diazotized polyaminopolystyrene resin. On the other hand, prior to this (1949), immobilization of physiologically active protein was carried out by Micheel and Ewers.[4] Further, in 1951, Campbell et al.[5] prepared immobilized antigen by binding albumin to diazotized p-aminobenzylcellulose. Subsequently, a number of reports on the preparation and application of immobilized antigens and antibodies appeared. Such studies may also be applicable to immobilized enzymes. Following Grubhofer's investigation, less than ten papers were published on immobilized enzymes in the 1950's. In the 1960's, many papers appeared; in particular, Professor Katzir-Katchalski and his co-workers at the Weizmann Institute of Science, Israel, carried out extensive studies on new immobilization techniques, and on the enzymatic, physical and chemical properties of immobilized enzymes.

Since the early 1960's, the present authors have investigated immobilized enzymes with the aim of utilizing them for continuous industrial production. The first report on immobilized aminoacylase was presented at the annual meeting of the Agricultural Chemical Society of Japan in 1965, and published in Enzymologia in 1966. In 1969, the authors succeeded in the industrialization of a continuous process for the optical resolution of DL-amino acids using immobilized aminoacylase. This was the first industrial application of immobilized enzymes in the world.

In the late 1960's, work on immobilized enzymes was carried out extensively in the U.S.A. and Europe, and the number of reports on immobilized enzymes increased markedly, as shown in Fig. 1.1. Besides these reports, many reviews[6-20] and books[21-33] have been also published.

In 1971, the first Enzyme Engineering Conference was held at Henniker, New Hampshire, U.S.A., and the predominant theme of this conference was immobilized enzymes. As will be described later, a definition and classification of immobilized enzymes were proposed at the conference. This conference is biannual, and the main topics have continued to be immobilized enzymes. In Japan, since the end of the 1960's, considerable work has appeared on immobilized enzymes, and a

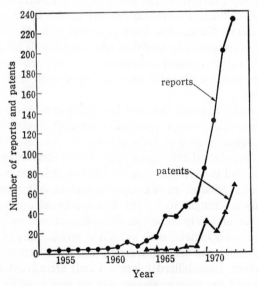

Fig. 1.1   Trends in studies of immobilized enzymes.
Redrawing from *Kagaku to Seibutsu* (Japanese), **15** (1), 48 (1977), fig. 1

symposium on immobilized enzymes was held at the 4th International Fermentation Symposium at Kyoto in March, 1972. Professor Katzir-Katchalski from Israel and Dr. Chibata, editor of this book, from Japan chaired the symposium, and there were extensive and fruitful discussions. Since then, studies on immobilized enzymes have continued at a rapid pace, and at present Japan is one of the leading countries in the world in this field.

Recently, applications of immobilized enzymes have expanded to new fields in addition to chemical synthesis. In the late 1960's, immobilization of physiologically active substances, including enzymes, was successfully carried out by Professor Porath and his co-workers at Uppsala University, Sweden. These immobilization techniques were developed to produce a specific isolation procedure for biological substances, termed affinity chromatography, by Professor Anfinsen of NIH, U.S.A., and Professor Cuatrecasas of Johnes Hopkins University, U.S.A.

Although enzymes are produced by all living things, enzymes from microbial sources are most suitable for industrial purposes for the following reasons: 1) the production cost is low, 2) conditions for production are not restricted by location and season, 3) the time required for production is short, and 4) mass production is possible. Microbial enzymes can be classified into two groups. One consists of enzymes excreted from the cells into the growth medium, and the other consists of enzymes retained in the cells during cultivation. The former type is called extracellular and the latter intra-cellular. For the utilization of the latter enzymes, it is necessary to extract the enzyme from microbial cells. However, such extracted enzymes are generally unstable, and not suitable for practical use. In recent years, many chemical substances have been produced by fermentation methods utilizing the catalytic activities of multi-enzyme systems in microorganisms.

Thus, in order to avoid the need to extract enzymes from microbial cells or to utilize multi-enzyme systems of the cells, direct immobilization of whole microbial cells has been attempted. As described later, the authors have investigated continuous enzyme reactions using immobilized microbial cells, and in 1973 succeeded in the industrialization of a process for the continuous production of L-aspartic acid using immobilized microbial cells. This is thought to be the first industrial application of immobilized microbial cells in the world. In the case of immobilized microbial cells, many problems remain, such as limited permeability of the substrate and product through the cell membranes, and the occurrence of side reactions. However, immobilized microbial cells are advantageous, as the enyzme systems of the microorganism can be readily utilized, and their future seems very promising.

1.2 DEFINITION OF IMMOBILIZED ENZYMES

Immobilized enzymes are defined as "enzymes physically confined or localized in a certain defined region of space with retention of their catalytic activities, and which can be used repeatedly and continuously". Accordingly, enzymes modified to a water-insoluble form by suitable techniques satisfy this definition of immobilized enzymes. In addition, when an enzyme reaction using a substrate of high molecular weight is carried out in a reactor equipped with a semipermeable ultrafiltration membrane, a reaction product of low molecular weight can be removed continuously through the membrane without leakage of the enzyme from the reactor. This also can be considered as a kind of immobilized enzyme system.

The term "immobilized enzyme" was recommended at the 1st Enzyme Engineering Conference in 1971. Before that time, various terms such as "water-insoluble enzyme", "trapped enzyme", "fixed enzyme" and "matrix-supported enzyme" had been used. In this conference, some problems relating to immobilized enzymes, including the terminology, were discussed. A classification of immobilized enzymes was proposed, as shown in Fig. 1.2. Enzymes are first classified into native enzymes and

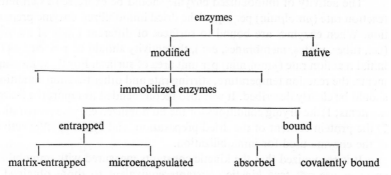

Fig. 1.2   Classification of immobilized enzymes.
From *Enzyme Eng.* (ed. L. B. Wingard), J. Wiley & Sons, 1972, fig. 1

modified enzymes. Immobilized enzymes belong to the latter type, which also includes chemically modified soluble enzymes and biologically, i.e., genetically, modified enzymes.

Thus, for practical use as catalysts, enzymes in the following three forms can be considered: 1) soluble form, 2) soluble immobilized form, and 3) insoluble immobilized form. For the latter two forms, the term "immobilized enzyme" is more suitable than "insoluble enzyme".

Although not fundamentally different from the above classification, in this book immobilization of enzymes is classified into "carrier-binding", "cross-linking" and "entrapping" types, as shown in Fig. 1.3. Carrier-

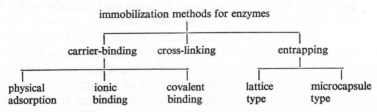

Fig. 1.3 Classification of immobilization methods for enzymes.

binding is subdivided into "physical adsorption", "ionic binding" and "covalent binding", while entrapping is divided into "lattice type" and "microcapsule type". As described later, this classification is reasonable and convenient. Except for the case using an ultrafiltration membrane, immobilized enzymes are apparently insoluble in water, and the enzyme reaction is carried out in a heterogenous medium. Since this introduces a number of complications, definitions and expressions of kinetic parameters for immobilized enzymes were also recommended by the Enzyme Engineering Conference. The recommendations can be outlined as follows.

The activity of immobilized enzyme should be expressed as an initial reaction rate (μmol/min) per mg of the dried immobilized enzyme preparation. When enzymes are bound to surfaces of different kinds of matrices (i.e., tubes, plates, membranes, etc.), the activity should be reported as the initial reaction rate (μmol/min) per unit area of surface. For these measurements, the reaction temperature, stirring rate and other reaction conditions should be clearly described. It was also recommended to report the following items; 1) the drying conditions for the immobilized enzyme preparation, 2) the protein content of the dried preparation, and 3) the specific activity of the enzyme used for immobilization.

It is recognized that the kinetic constants measured with immobilized enzymes are not true kinetic constants equivalent to those obtained in homogeneous reactions, but are apparent values because of the effects of diffusion and other physical factors. Hence, maximum velocity and Michaelis constants should be referred to as apparent V_{max} and apparent K_m (V_{max} (app) and K_m (app)). Other kinetic constants should also be reported as the apparent constants.

When reporting on the stability of enzyme activity of an immobilized preparation, it is strongly recommended to describe in detail the specific conditions used for the stability measurements.

When a new carrier is developed for the immobilization of enzymes, it is desirable to describe its properties and characteristics, such as the number of reactive groups per unit weight of carrier and the maximum binding amount of a small molecule as well as a large molecule (protein) to a specified weight of the carrier. These recommendations are very useful as a unifying factor for the future development of immobilized enzymes.

1.3 DEFINITION OF IMMOBILIZED MICROBIAL CELLS

Immobilized microbial cells can be defined by replacing the word enzymes with microbial cells in the definition given for immobilized enzymes. That is, "microbial cells physically confined or localized in a certain defined region of space with retention of their catalytic activities, and which can be used repeatedly and continuously." In this immobilized cells, the cells are in growing, resting and/or autolyzed states. In some cases, the immobilized microbial cells are dead, but the enzyme activities remain. When cells are growing state, it is sometimes difficult to differentiate immobilized systems as defined above from certain kinds of continuous fermentation processes.

The nomenclature of enzymes used in this book follows recommended names or other names described in *Comprehensive Biochemistry* (edited by M. Florkin and E.H. Stotz, Elsevier, 1973). For enzymes not described there, the nomenclature of the original report is used.

REFERENCES

1. J. M. Nelson and E. G. Griffin, *J. Am. Chem. Soc.*, **38**, 1109 (1916)
2. J. B. Sumner, *Science*, **108**, 410 (1948)
3. N. Grubhofer and L. Schleith, *Naturwissenschaften*, **40**, 508 (1953)
4. F. Micheel and J. Ewers, *Makromol. Chem.*, **3**, 200 (1949)
5. D. H. Campbell, E. Luescher and L. S. Lerman, *Proc. Nat. Acad. Sci.*, **37**, 575 (1951)
6. E. Katchalski, *Polyamino Acids, Polypeptides and Proteins*, p. 283, The Univ. of Wisconsin Press, 1962
7. G. Manecke, *Naturwissenschaften*, **51**, 25 (1964)
8. G. G. Guilbault, *Enzymatic Method of Analysis*, p. 235, Pergamon Press, 1969
9. L. Goldstein, *Fermentation Advances*, p. 391, Academic Press, 1969
10. A. S. Lindsey, *Reviews in Macromolecular Chemistry*, vol. 4, p. 1, Marcel Dekker, 1970
11. L. Goldstein, *Method in Enzymology*, vol. 19, p. 935, Academic Press, 1970
12. E. Katchalski, I. Silman, and R. Goldman, *Advances in Enzymology*, vol. 34, p. 445, John Wiley & Sons, 1971
13. F. Friedberg, *Chromatog. Rev.*, **14**, 121 (1971)
14. G. J. H. Melrose, *Rev. Pure Appl. Chem.*, **21**, 83 (1971)
15. K. L. Smiley and G. W. Strandberg, *Advances in Applied Microbiology*, vol. 15, p. 13, Academic Press, 1972
16. S. Aiba, A. E. Humphrey, and N. F. Millis, *Biochemical Engineering*, p. 393, Univ. of Tokyo Press, 1973

17. K. F. O'Driscoll, *Advances in Biochemical Engineering*, vol. 4, p. 155, Springer-Verlag, 1976
18. K. Mosbach, *Techniques of Chemistry*, vol. 10, Part 2, p. 969, John Wiley & Sons, 1976
19. N. Lakshminarayanaiah, *Membrane Electrodes*, p. 335, Academic Press, 1976
20. T. R. Jack and J. E. Zajic, *Advances in Biochemical Engineering*, vol. 5, p. 125, Springer-Verlag, 1977
21. G. R. Stark, *Biochemical Aspects of Reactions on Solid Supports*, Academic Press, 1971
22. L. B. Wingard, Jr. (ed), *Enzyme Engineering*, John Wiley & Sons, 1972
23. O. R. Zaborsky, *Immobilized Enzymes*, Chemical Rubber Co. Press, 1972
24. A. C. Olson and C. L. Cooney (ed), *Immobilized Enzymes in Food and Microbial Processes*, Plenum Press, 1974
25. M. Salmona, C. Saronio, and S. Garattini (ed), *Insolubilized Enzymes*, Raven Press, 1974
26. R. B. Dunlap (ed), *Advances in Experimental Medicine and Biology*, vol. 42, Plenum Press, 1974
27. S. J. Gutcho, *Immobilized Enzymes*, Noyes Data Corporation, 1974
28. H. H. Weetall (ed), *Immobilized Enzymes, Antigens, Antibodies, and Peptides*, Marcel Dekker, 1975
29. H. H. Weetall and S. Suzuki (ed), *Immobilized Enzyme Technology*, Plenum Press, 1975
30. R. A. Messing, *Immobilized Enzymes for Industrial Reactors*, Academic Press, 1975
31. K. Mosbach (ed), *Methods in Enzymology*, vol. 44, Academic Press, 1976
32. L. B. Wingard, Jr., E. Katchalski-Katzir, and L. Goldstein (ed), *Applied Biochemistry and Bioengineering*, vol. 1, Academic Press, 1976
33. T. M. S. Chang (ed), *Biomedical Application of Immobilized Enzymes and Proteins*, vol. 1, Plenum Press, 1977
34. L. B. Wingard, Jr. (ed), *Enzyme Engineering*, p. 16, John Wiley & Sons, 1972

CHAPTER 2

Preparation of immobilized enzymes and microbial cells

2.1 IMMOBILIZATION METHODS FOR ENZYMES

A specific conformation and an active center interacting with the substrate are regarded as essential features of the catalytic activity of enzymes. The active center consists of two sites having different functions. One is the reactive or catalytic site participating in the catalytic action, and the other is the specific or binding site controlling the substrate specificity of the enzyme. These sites are usually composed of several amino acid residues held in a specific spatial relationship. The three-dimensional conformation of the entire enzyme protein also has an important effect on the catalytic activity. Consequently, to retain the catalytic activity of the enzyme in the immobilized state, it is necessary to retain the native structure as far as possible. Accordingly, in order to prepare active immobilized enzymes, immobilization should be carried out under very mild and extremely well-controlled conditions. If the amino acid residues at the active center, or the tertiary structure, are altered, the catalytic activity may decrease and changes of enzymatic properties such as substrate specificity may occur.

As functional groups involved in enzyme immobilization, free amino and carboxyl groups, the sulfhydryl group of cysteine, the imidazole group of histidine, phenolic groups, and hydroxyl groups of serine and threonine may be considered. For immobilization of an enzyme it is necessary that functional groups in the active center should not be involved in the reaction leading to immobilization of the enzyme. Further, since the tertiary structure of enzyme protein is maintained by relatively weak binding forces, such as hydrogen, hydrophobic and ionic bonds, it is necessary to carry out the immobilization reaction under mild conditions, as already mentioned. Therefore, reactions at high temperature, and strong acid or alkali treatments must be avoided to preserve the structural integrity of enzymes. Even treatments with organic solvents or high salt concentrations may cause denaturation and loss of activity.

9

Methods for enzyme immobilization can be classified into three basic categories as follows.

1) Carrier-binding method: the binding of enzymes to water-insoluble carriers.

2) Cross-linking method: intermolecular cross-linking of enzymes by means of bifunctional or multifunctional reagents.

3) Entrapping method: incorporating enzymes into the lattice of a semipermeable gel or enclosing the enzymes in a semipermeable polymer membrane. These methods are shown schematically in Fig. 2.1.

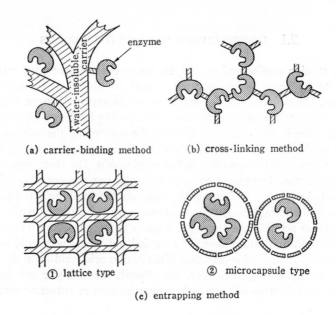

(a) carrier-binding method   (b) cross-linking method

① lattice type   ② microcapsule type

(c) entrapping method

Fig. 2.1 Schematic diagrams of immobilized enzymes.

2.1.1 Carrier-binding Method

The carrier-binding method is the oldest immobilization method for enzymes and many papers have been published on this method. When enzymes are immobilized in this way, care is required regarding the selection of carriers as well as in binding techniques. Namely, the amount of enzyme bound to the carrier and the activity after immobilization depend markedly on the nature of the carrier. Although the selection of a carrier also depends on the nature of the enzyme itself, the following aspects must be considered; (1) particle size, (2) surface area, (3) molar ratio of hydrophilic to hydrophobic groups, and (4) chemical composition.

Generally speaking, increases in the ratio of hydrophilic groups and in surface area increase the amount of bound enzyme per unit carrier, resulting in higher activity of the immobilized enzyme.

As carriers for enzyme immobilization, polysaccharide derivatives such as cellulose, dextran, agarose and polyacrylamide gel are most commonly used. Some are available commercially under trade names such as Sephadex, Sepharose, Bio-Gel, and so forth. A porous glass developed by Corning Glass Works, U.S.A., has also been employed as a carrier for enzyme immobilization.

The carrier-binding method can be further divided into three categories according to the binding mode of the enzyme, that is, physical adsorption, ionic binding and covalent binding.

A. Physical adsorption method

The physical adsorption method for the immobilization of enzyme is based on the physical adsorption of enzyme protein on the surface of water-insoluble carriers. This method often causes little or no conformational change of the enzyme protein, or destruction of its active center. Accordingly, if a suitable carrier for the target enzyme is found, this method may be both simple and effective. However, this method has the disadvantage that the adsorbed enzyme may leak from the carrier during utilization, because the binding force between the enzyme protein and carrier is weak.

Immobilized enzymes which have been prepared by physical adsorption are listed in Table 2.1.

As carriers for this method, inorganic materials such as activated carbon, porous glass, acid clay, bleaching clay, kaolinite, alumina, silica gel, bentonite, hydroxylapatite and calcium phosphate gel, as well as natural polymers such as starch and gluten, have been employed. Among these carriers, activated carbon has a long history, and a number of papers have appeared on its use.

As described previously (Section 1.1), Nelson and Griffin[15] observed in 1916 that invertase adsorbed on activated carbon retained its activity toward saccharose, though it was not their intention to immobilize the enzyme. An example of immobilization using activated carbon will be given here.[9] Activated carbon was added to α-amylase solution and the mixture was stirred at 10°C for 1 h. After filtration the resulting α-amylase-activated carbon complex was packed into a column. On feeding starch solution into this enzyme column, conversion of starch to glucose occurred.

TABLE 2.1   Immobilization of enzymes by physical adsorption

Carrier	Enzyme	Reference
ORGANIC SUPPORTS		
Concanavalin A	arylsulfatase	1
Gluten	β-amylase	2
Tannin-aminohexyl cellulose	naringinase	3
	lactase (β-galactosidase)	3
	aminoacylase	3, 4
	papain	3
	glucose isomerase	3
Concanavalin A–Sepharose	phosphodiesterase	5
Butyl–Sepharose	lipoamide dehydrogenase	6
Hexyl–Sepharose	lipoamide dehydrogenase	6
Starch	α-amylase	7
Activated carbon	glucose oxidase	8
	α-amylase	9
	β-amylase	10
	glucoamylase	8, 10–14
	invertase	15, 16
INORGANIC SUPPORTS		
Alumina	glucose oxidase	8
	glucoamylase	8, 17
Bentonite	invertase	18
	cephalosporin acetylesterase	19
Calcium phosphate gel	leucine aminopeptidase	20
	aspartase	21
Clay (acid)	glucose oxidase	8
	transglucosidase	22
	glucoamylase	8, 13, 23
Clay (bleaching)	α-amylase	9
Hydroxylapatite	NAD pyrophosphorylase	24
	leucine aminopeptidase	20
Kaolinite	lysozyme	25
	invertase	26
Nickel–silica–alumina	catalase	27
Porousglass	glucose oxidase	28
	ribonuclease	29, 30
	lipase	31
	papain	28, 32
Silica gel	phospholipase	33
	aspartase	21
Titania	glucose oxidase	34
Titanium oxide–stainless steel	lactase	35

Further, immobilization of an enzyme by hydrophobic binding to carriers was reported.[6] Namely, lipoamide dehydrogenase was immobilized by binding to carriers containing hydrophobic residues, such as butyl- or hexyl-Sepharose.

More recently, we have found that enzymes can be readily immobilized by using an adsorbent containing tannin as a ligand.[3,4,36] Namely, the adsorbent is prepared by the reaction of tannin previously activated by CNBr or epichlorohydrin with aminohexyl cellulose. The tannin-aminohexyl cellulose can be used for immobilization of many kinds of enzymes; hydrophobic forces play an important role in this immobilization.

B. Ionic binding method

The ionic binding method is based on the ionic binding of enzyme protein to water-insoluble carriers containing ion-exchange residues. In some cases, not only ionic binding but also physical adsorption may take part in the binding. Consequently, in this section, cases in which ionic binding appears to play a more important role than physical adsorption are described.

As carriers for ionic binding, polysaccharides and synthetic polymers having ion-exchange residues are used. The binding of enzyme to the carrier is easily carried out, and the conditions are mild in comparison with those necessary for the covalent binding method. Therefore, the ionic binding method causes little or no changes of conformation and the active site of the enzyme protein, and yields immobilized enzymes having high activity in many cases. As the binding forces between enzyme protein and carriers are less strong than in covalent binding, leakage of enzyme from the carrier may occur in substrate solutions of high ionic strength or upon variation of pH. Immobilized enzymes prepared by the ionic binding method are listed in Table 2.2.

Immobilization of an enzyme by this method was first reported by Mitz[39] in 1956. He prepared immobilized catalase by passing a catalase solution through a column packed with DEAE-cellulose. When hydrogen peroxide solution was passed through the column, no hydrogen peroxide was detected in the effluent.

In order to produce L-amino acids efficiently from chemically synthesized DL-amino acids, we studied the immobilization of mold aminoacylase[47] in detail. Highly active and stable immobilized aminoacylases were obtained by using anion-exchange carriers such as DEAE-cellulose, TEAE-cellulose, ECTEOLA-cellulose and DEAE-Sephadex. For example, DEAE-Sephadex (OH form) was added to aminoacylase solution dissolved in 0.1 M phosphate buffer (pH 7.0), and the mixture was stirred at 37°C for 5 h. This simple procedure caused aminoacylase to become ionically bound to the DEAE-Sephadex.

Further, ionic binding to carriers has been reported after increasing the charge on the enzyme protein by chemical modification.[42] Namely, slightly anionic glucoamylase was modified to a polyanionic derivative by

TABLE 2.2   Immobilization of enzymes by ionic binding

Carrier	Enzyme	Reference
ANION EXCHANGERS		
DEAA-cellulose	invertase	37
DEAE-cellulose	catalase	38, 39
	glucoamylase	40–42
	invertase	16, 43, 44
	pepsin	45
	protease	45
	asparaginase	46
	aminoacylase	47–49
	aminoacylase I (kidney)	50
	ATP deaminase	51
	AMP deaminase	51
	inorganic pyrophosphatase	24
	aspartase	21
ECTEOLA-cellulose	aminoacylase	47
	D-oxynitrilase	52
	aspartase	21
TEAE-cellulose	aminoacylase	47
	ATP deaminase	51
	aspartase	21
DEAE–Sephadex	dextransucrase	53
	aminoacylase	47, 54–56
	α-amino-ε-caprolactam hydrolase	57
	aspartase	21
	α-amino-ε-caprolactam racemase	57
	glucose phosphate isomerase	58
n-Octylamino–Sephadex	lactate dehydrogenase	59
	xanthine oxidase	59
	ribonuclease	59
	alkaline phosphatase	59
	urease	59
Polyaminopolystyrene	catalase	60
Amberlite IR-45	glucoamylase	61
	invertase	62
Amberlite IRA-93	invertase	62
Amberlite IRA-410	invertase	62
Amberlite IRA-900	invertase	62
CATION EXCHANGERS		
CM–cellulose	trypsin	45
	α-chymotrypsin	45
	asparaginase	63
	invertase	64
Cellulose–citrate	trypsin	45
	α-chymotrypsin	45

(*Continued*)

TABLE 2.2—*Continued*

Carrier	Enzyme	Reference
P-cellulose	trypsin	60
	α-chymotrypsin	60
Amberlite CG-50	glucose oxidase	8, 65
	glucoamylase	8
Amberlite IRC-50	invertase	62
Amberlite IR-120	invertase	62
Amberlite IR-200	invertase	62
Amberlite XE-97	lipase	60
Dowex-50	ribonuclease	66

covalent binding with a water-soluble copolymer of ethylene and maleic anhydride. The polyanionic enzyme derivative was strongly bound to cationic carriers such as DEAE-cellulose and DEAE-Sephadex. The resulting immobilized glucoamylase was much more stable than the immobilized preparation obtained using untreated enzyme.

C. Covalent binding method

The covalent binding method is based on the binding of enzymes and water-insoluble carriers by covalent bonds. Among the various carrier-binding methods, most studies have utilized this method.

The functional groups that take part in the covalent binding of enzyme to carrier are as follows. (1) α- or ε-amino group, (2) α-, β-, or γ-carboxyl group, (3) sulfhydryl group, (4) hydroxyl group, (5) imidazole group, (6) phenolic group. In coupling reactions, these functional groups react with carriers containing reactive groups such as diazonium, acid azide, isocyanate and halides.

This method can be further classified into diazo, peptide and alkylation methods according to the mode of linkage. The selection of conditions for immobilization by covalent binding is more difficult than in the cases of physical adsorption and ionic binding. The reaction conditions required for covalent binding are relatively complicated and not particularly mild. Therefore, in some cases, covalent binding alters the conformational structure and active center of the enzyme, resulting in major loss of activity and/or changes of substrate specificity. However, the binding force between enzyme and carrier is strong and leakage of the enzyme does not occur even in the presence of substrate or salt solutions of high ionic strength.

a. Diazo Method

The diazo method is based on the diazo coupling of enzyme proteins and diazonium derivatives of water-insoluble carriers. That is, carriers containing aromatic amino groups are diazotized with nitrous acid, and then the diazonium derivatives are reacted with enzyme proteins as shown in Eq. 2.1.

$$R–Y–NH_2 \xrightarrow{NaNO_2/HCl} (R–Y–N^+ \equiv N)Cl^- \xrightarrow{enzyme} R–Y–N=N\text{-enzyme} \qquad (2.1)$$

The functional groups on enzyme proteins participating in diazo coupling include free amino group, the imidazole group of histidine and the phenolic group of tyrosine. Immobilized enzymes which have been prepared by this method are listed in Table 2.3.

TABLE 2.3   Immobilization of enzymes by the diazo method

Carrier	Enzyme	Reference
POLYSACCHARIDE DERIVATIVES		
m-Aminoanisole cellulose	catalase	38
	ribonuclease	38
	trypsin	38
	α-chymotrypsin	38
m-Aminobenzyloxymethyl cellulose	trypsin	38
	α-chymotrypsin	38
	thrombin	67
p-Aminobenzyl cellulose	catalase	38
	creatine kinase	68
	ribonuclease	38, 69, 70
	ribonuclease T_1	71–73
	α-amylase	74
	invertase	16
	lysozyme	75
	trypsin	38, 69
	α-chymotrypsin	38, 69, 70, 76
	papain	77
	aminoacylase	78–80
	myosin ATPase	77
	aldolase	81
	glutamate dehydrogenase	82
	aspartase	21
p-Aminobenzoyl cellulose	trypsin	38
	α-chymotrypsin	38
3-(p-Amino–m-methyl-anilino)–5-chlorotriazinyl cellulose	trypsin	38
3-(p-Aminophenoxy)–2-hydroxypropyl cellulose	α-amylase	83
	β-amylase	83
Sephadex–anthranyl ester	trypsin	84
	subtilopeptidase A	84

(*Continued*)

TABLE 2.3—*Continued (1)*

Carrier	Enzyme	Reference
	subtilopeptidase B	84
Starch–methylenedianiline	trypsin	85
compound	papain	85
	subtilopeptidase A	85
AMINO ACID COPOLYMER		
p-Amino–DL-phenylalanine–	ribonuclease	86, 87
L-leucine	pepsin	88
	trypsin	89–94
	α-chymotrypsin	88, 89
	papain	94–97
	prothrombin	98
	streptokinase	99, 100
	pronase	101
	urease	102
POLYACRYLAMIDE DERIVATIVES		
Bio-Gel P-2 aminoethyl	lysozyme	75
Enzacryl AA	α-amylase	83, 103
	β-amylase	83, 103
	aminoacylase	80
	aminoacylase I (kidney)	104
STYRENE RESINS		
Polyaminopolystyrene	catalase	60, 105
	ribonuclease	106, 107
	carboxypeptidase	106, 107
	lipase	60
	α-amylase	108
	β-fructofuranosidase	109
	diastase	106, 107
	pepsin	106, 107
	aminoacylase	69, 70, 110
	apyrase	111
	L-glutamate dehydrase	82
Copolymer of methacrylate–	papain	112
m-aminostyrene		
COPOLYMER OF ETHYLENE AND MALEIC ACID		
p,p'-Diaminodiphenyl	trypsin	113
derivative	α-chymotrypsin	113
	papain	113
	subtilisin	113
AMINOSILANE DERIVATIVES		
Aminosilanized porous glass	lactate dehydrogenase	114–116
	glucose oxidase	117, 118
	L-amino acid oxidase	119, 120
	uricase	121

(*Continued*)

TABLE 2.3—*Continued* (2)

Carrier	Enzyme	Reference
	hydrogenase	122
	steroid esterase	123
	pyruvate kinase	115
	acetylcholinesterase	124
	alkaline phosphatase	125
	deoxyribonuclease I	126
	glucoamylase (amyloglucosidase)	127
	lactase (β-galactosidase)	128–132
	invertase (saccharase)	133
	leucine aminopeptidase	134
	aminopeptidase M	134
	trypsin	135–139
	α-chymotrypsin	140
	papain	135, 136, 141
	ficin	135
	pronase	142
	neutral protease	143
	urease	144
	aminoacylase	80
	ribulose diphosphate carboxylase	145
	glucose isomerase	146
Aminosilanized Dacron	asparaginase	147

As carriers for this method, aromatic amino derivatives of polysaccharides, copolymers of amino acids, polyacrylamide, polystyrene, copolymers of ethylene-maleic acid and porous glass are commonly employed.

i) Polysaccharide derivatives

Polysaccharide derivatives containing aromatic amino groups include p-aminobenzyl cellulose (1), m-aminoanisole cellulose (2), m-aminobenzyl-oxymethyl cellulose (3), p-aminobenzoyl cellulose (4), 3-(p-aminophenoxy)-2-hydroxypropionyl cellulose (5), 3-(p-amino-m-methylanilino)-5-chloro-triazinyl cellulose (6), Sephadex-anthranylester (7), starch-methylene dianiline compound (8), and so on. The cellulose derivatives in particular have been used by many investigators and for many different enzymes.

p-Aminobenzyl cellulose is most commonly employed. This carrier was first used by Campbell et al.,[148] who used the diazonium derivative to immobilize bovine serum albumin as an antigen and used it for the isolation of specific antibody from rabbit serum. Since then, this cellulose derivative has been used for the immobilization of many enzymes. Starch-methylene dianiline derivative (8), which was prepared by the condensation

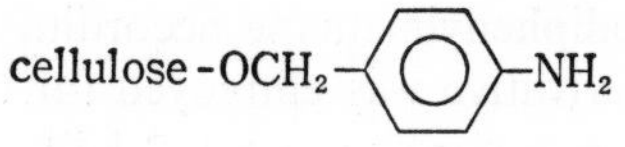
cellulose-OCH$_2$
NH$_2$

1

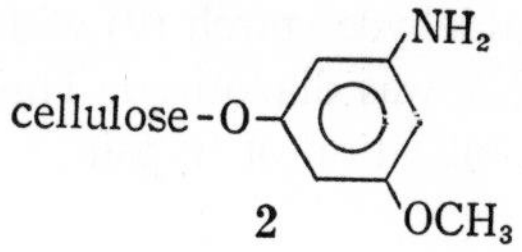
cellulose-O
NH$_2$
OCH$_3$

2

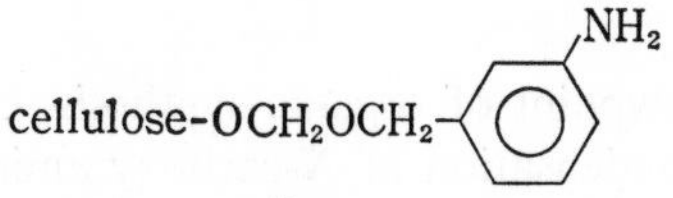
cellulose-OCH$_2$OCH$_2$
NH$_2$

3

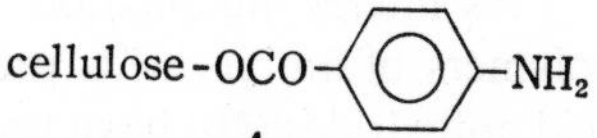
cellulose-OCO
NH$_2$

4

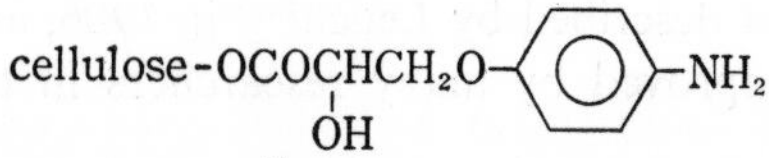
cellulose-OCOCHCH$_2$O
OH
NH$_2$

5

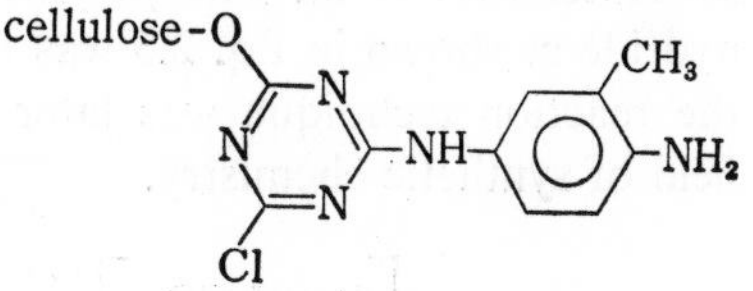
cellulose-O
N
N
N
Cl
NH
CH$_3$
NH$_2$

6

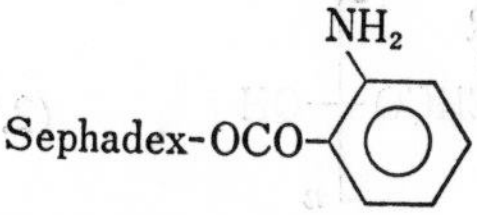
Sephadex-OCO
NH$_2$

7

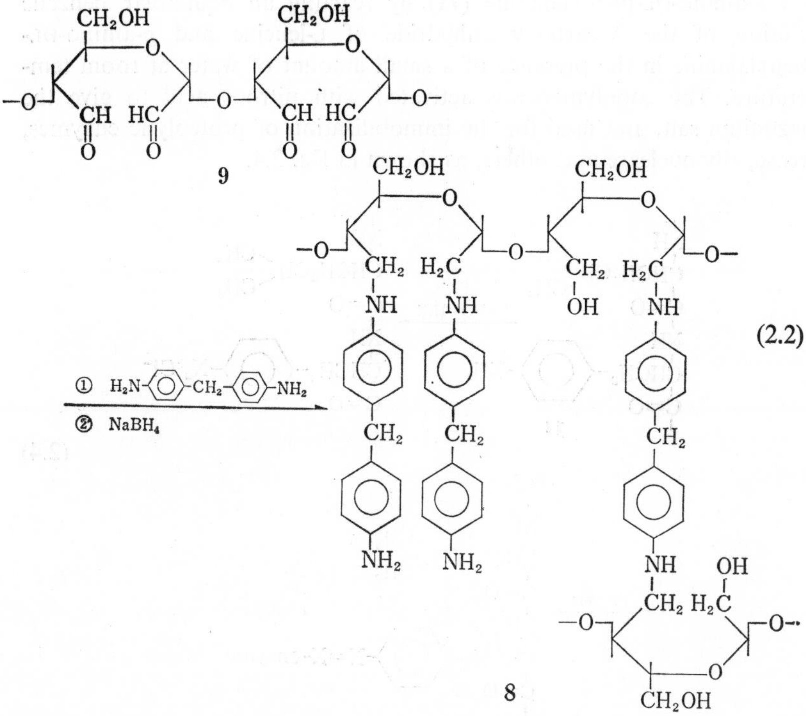
CH$_2$OH
CH
HC
O
① H$_2$N-⟨⟩-CH$_2$-⟨⟩-NH$_2$
② NaBH$_4$
CH$_2$
H$_2$C
NH
OH
NH$_2$

9

(2.2)

8

of dialdehyde starch (9) with p,p'-diaminodiphenylmethane according to Eq. 2.2, was diazotized. This diazotized derivative was employed for the immobilization of trypsin, papain and others according to Eq. 2.1.[85]

ii) Copolymers of amino acids

As interesting method from the viewpoint of organic synthesis, copolymers of amino acids prepared by condensation of *N*-carboxy amino acid anhydrides (10) have been applied for the immobilization of enzymes by Katchalski *et al.* The polymerization of *N*-carboxy amino acid anhydride as shown in Eq. 2.3 was first described by Leuch[149] in 1906, and the reaction technique was later improved by many researchers in the field of synthetic chemistry.

$$n \begin{bmatrix} R\text{-}CH\text{-}CO \\ \quad \big| \quad \searrow O \\ NH\text{-}CO \end{bmatrix} \longrightarrow H \begin{bmatrix} R \\ \big| \\ \text{-}NHCHCO\text{-} \end{bmatrix}_n OH \qquad (2.3)$$

10

For instance, Katchalski *et al.*[88-96,102] prepared a copolymer of L-leucine and *p*-amino-DL-phenylalanine (11) by reacting an equimolar benzene solution of the *N*-carboxy anhydride of L-leucine and *p*-amino-DL-phenylalanine in the presence of a small amount of water at room temperature. The copolymer was activated with nitrous acid to give the diazonium salt, and used for the immobilization of proteolytic enzymes, urease, ribonuclease and others, as shown in Eq. 2.4.

$$(2.4)$$

For example, immobilized urease having higher activity was obtained by stirring a mixture of urease solution and the diazotized copolymer at 4°C for 20 h.

iii) Polyacrylamide derivatives

Polyacrylamide derivatives are available commercially as Bio-Gel or Enzacryl, and these aromatic amino derivatives can be coupled by the diazo method to immobilize enzymes. For example, Enzacryl AA is a polyacrylamide derivative containing aromatic amino groups (**12**), and after diazotization it has been used for the immobilization of aminoacylase, α-amylase and others according to Eq. 2.5.

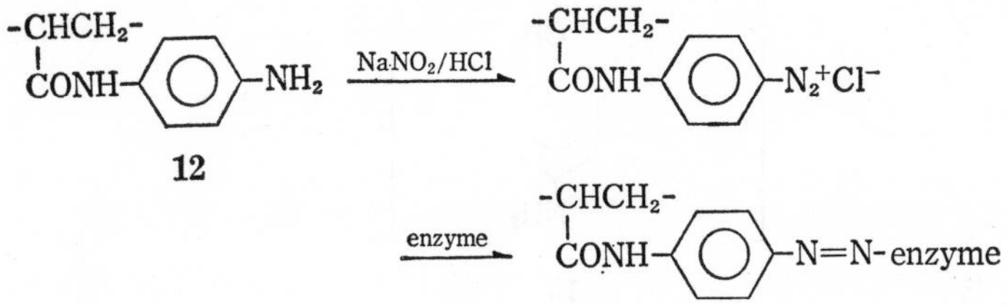

$$(2.5)$$

However, the carriers are relatively expensive and there have been few reports of immobilization of enzymes.

iv) Styrene resins

Polyaminopolystyrene (**13**) and a copolymer of methacrylate-*m*-aminostyrene (**14**) have been used for the immobilization of enzymes.

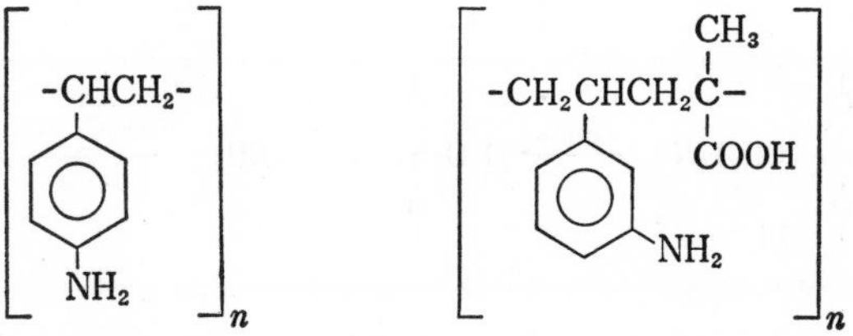

13   14

Grubhofer *at al.*[106] first coupled pepsin, ribonuclease and other enzymes to diazotized polyaminopolystyrene according to Eq. 2.1 and carried out continuous enzyme reactions using columns packed with immobilized enzymes.

v) Copolymer of ethylene and maleic acid

Although this copolymer is described in detail in the next section on the peptide binding method, its diaminodiphenylmethane derivative (**15**) has been used as a carrier for immobilization of enzymes by the diazo method.

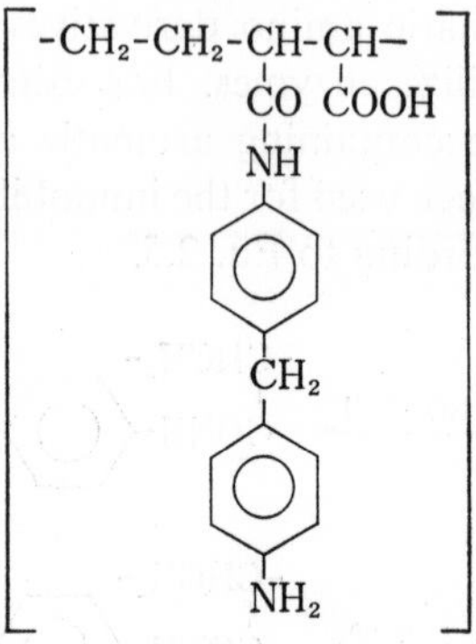

$$
\left[
\begin{array}{c}
-CH_2-CH_2-CH-CH- \\
\underset{|}{CO} \quad \underset{|}{COOH} \\
NH \\
\bigcirc \\
CH_2 \\
\bigcirc \\
NH_2
\end{array}
\right]_n
$$

15

vi) Aminosilane derivatives of porous glass

Aromatic aminosilane derivatives of porous glass were developed as a carrier for the immobilization of enzymes by Weetall *et al.* (Corning Glass Works), and diazotized derivatives have been used for the immobilization of quite a large number of enzymes.

As shown in Eq. 2.6, aminoalkyl groups are incorporated into porous

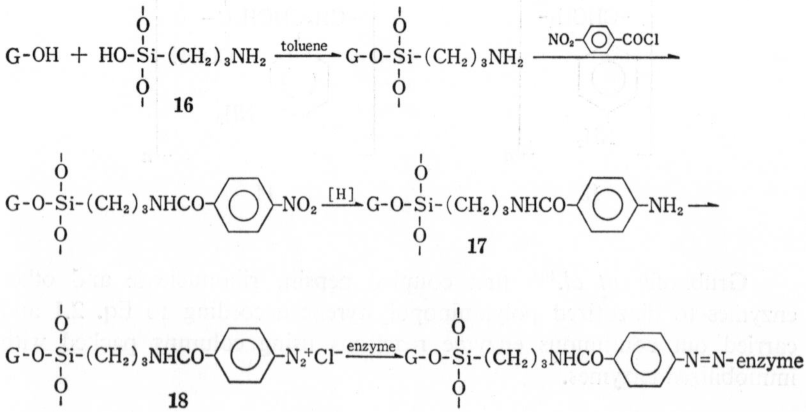

$$
G\text{-OH} + HO\text{-}\underset{O}{\overset{O}{Si}}\text{-}(CH_2)_3NH_2 \xrightarrow{\text{toluene}} G\text{-}O\text{-}\underset{O}{\overset{O}{Si}}\text{-}(CH_2)_3NH_2 \xrightarrow{NO_2\text{-}\bigcirc\text{-}COCl}
$$

16

$$
G\text{-}O\text{-}\underset{O}{\overset{O}{Si}}\text{-}(CH_2)_3NHCO\text{-}\bigcirc\text{-}NO_2 \xrightarrow{[H]} G\text{-}O\text{-}\underset{O}{\overset{O}{Si}}\text{-}(CH_2)_3NHCO\text{-}\bigcirc\text{-}NH_2 \longrightarrow
$$

17

$$
G\text{-}O\text{-}\underset{O}{\overset{O}{Si}}\text{-}(CH_2)_3NHCO\text{-}\bigcirc\text{-}N_2^+Cl^- \xrightarrow{\text{enzyme}} G\text{-}O\text{-}\underset{O}{\overset{O}{Si}}\text{-}(CH_2)_3NHCO\text{-}\bigcirc\text{-}N=N\text{-enzyme}
$$

18

$$(2.6)$$

glass (*G*-OH) by refluxing porous glass with γ-aminopropyltriethoxysilane (16) in toluene. The resulting aminoalkylated porous glass is reacted with *p*-nitrobenzoylchloride, and then reduced to give the *p*-aminobenzoyl derivative (17). This is diazotized to the diazonium derivative (18), and coupled with enzyme as shown in Eq. 2.1.[125] This aromatic aminosilane-porous glass is sold by Corning Glass Works.

b. Peptide Binding Method

This binding method is an application of the peptide synthesis technique, and it is based on the formation of peptide bonds between the enzyme protein and a water-insoluble carrier.

The following procedures are available.

1) Water-insoluble carriers containing carboxyl groups can be converted to reactive derivatives such as the acid azide, chloride, isocyanate, etc., then these derivatives are reacted with free amino groups in the enzyme to form peptide bonds.

2) By using condensing reagents for peptide synthesis, such as carbodiimide and Woodward's reagent K, peptide bonds are formed between free carboxyl or amino groups in the enzyme and amino or carboxyl groups in the water-insoluble carrier.

Immobilized enzymes which have been prepared by the peptide binding method are listed in Table 2.4.

TABLE 2.4   Immobilization of enzymes by the peptide binding method

Carrier	Enzyme	Reference
ACID AZIDE DERIVATIVES		
CM-cellulose	lactate dehydrogenase	150
	glucose oxidase	135
	creatine kinase	68
	ribonuclease	70, 150, 151
	ribonuclease T_1	71–73, 152
	acetylcholinesterase	153
	α-amylase	74, 108
	glucoamylase	154, 155
	dextranase	156
	invertase	16
	trypsin	69,135,150,151,157
	α-chymotrypsin	69, 70,150,157,158
	papain	135, 150
	ficin	70,135,150,159,160
	bromelain	150, 161, 162
	streptokinase	163
	urokinase	163
	subtilisin	150

(*Continued*)

TABLE 2.4—*Continued (I)*

Carrier	Enzyme	Reference
	subtilopeptidase B	150
	pronase	150, 164
	ATPase	165, 166
	apyrase	153, 166, 167
	L-glutamate dehydrase	82
Copolymer of ethylene and	trypsin	113
maleic acid	α-chymotrypsin	113
	papain	113
	subtilisin	113
Copolymer of acrylamide and	trypsin	168
methacrylic acid	aminoacylase	168
Enzacryl AH	α-amylase	83, 103
	β-amylase	83
Phthaloyl-porous glass	pectin esterase	169
CARBOXYCHLORIDE RESINS		
Amberlite IRC-50	catalase	105, 170
	lipase	60
Amberlite XE-64	L-glutamate dehydrase	82
MALEIC ANHYDRIDE DERIVATIVES		
Copolymer of ethylene	glyceroaldehyde phosphate	81
and maleic anhydride	dehydrogenase	
	alkaline phosphatase	171
	naringinase	172
	trypsin	93, 135, 173–183
	α-chymotrypsin	86, 178, 181, 183
	papain	181, 183
	subtilisin	181
	pronase	164
	kallikrein	178
	apyrase	111
	aldolase	81
	fructosediphosphatase	81
Copolymer of butanediol	lactate dehydrogenase	150
divinylether and maleic	trypsin	150
anhydride	α-chymotrypsin	150
	papain	150
	ficin	150
	bromelain	150
	subtilisin	150
	subtilopeptidase B	150
	pronase	150
Copolymer of methyl vinyl	alkaline phosphatase	171
and maleic anhydride	naringinase	172
Copolymer of isobutyl vinyl	naringinase	172
ether and maleic anhydride		

(*Continued*)

TABLE 2.4—*Continued* (2)

Carrier	Enzyme	Reference
Copolymer of styrene and maleic anhydride	naringinase	172
ISOCYANATE DERIVATIVES		
CM-cellulose isocyanate derivative	ATPase	167
	apyrase	167
Sephadex isocyanate derivative	trypsin	184
	α-chymotrypsin	184
Polyaminopolystyrene iso-cyanate derivative	catalase	185
CNBr-ACTIVATED POLYSACCHARIDES		
CNBr-activated cellulose	xanthine oxidase	186
	β-D-glucosidase	187
	lactase	188
	polynucleotide phosphorylase	189, 190
	glucoamylase	191
	dextranase	157
	α-chymotrypsin	188
	pronase	164
	aminoacylase	80
CNBr-activated Sephadex	lactate dehydrogenase	192
	malate dehydrogenase	192
	glucose–6-phosphate dehydrogenase	193
	hexokinase	193, 194
	ribonuclease T_1	73
	lactase	193
	Taka-amylase A	195
	leucine aminopeptidase	196
	carboxypeptidase A	197
	trypsin	198, 199
	α-chymotrypsin	199–201
	aminoacylase	80
	citrate synthase	192
CNBr-activated Sepharose	lactate dehydrogenase	192, 202, 203
	malate dehydrogenase	192, 204
	isocitrate dehydrogenase	205
	glucose oxidase	206
	xanthine oxidase	186
	glutamate dehydrogenase	207
	D-amino acid oxidase	208
	pyridine nucleotide transhydrogenase	209
	peroxidase	210
	tryptophanase	202
	lipoxygenase	211
	protocatechuate-3,4-dioxy-genase	212

(*Continued*)

TABLE 2.4—*Continued* (3)

Carrier	Enzyme	Reference
	hepatic flavoprotein oxidase	213
	transaldolase	214
	phosphorylase	215, 216
	UDP-glucuronyltransferase	217
	aspartate aminotransferase	204
	tRNA-nucleotidyl transferase	218
	polynucleotide phosphorylase	189
	ribonuclease T_1	73
	glucoamylase	206, 219
	Taka-diastase A	195
	polygalacturonase	220
	mannosidase	221
	leucine aminopeptidase	196, 222
	aminopeptidase M	223
	carboxypeptidase C	224
	renin	225
	trypsin	198, 223, 226
	α-chymotrypsin	223, 227, 228
	elastase	229
	thrombokinase	230
	urokinase	231, 232
	subtilisin	233
	rennin	227
	neutral protease	234
	prolidase	223
	aldolase	235–237
	citrate synthase	192
	succinyl-CoA synthetase	238
CNBr-activated copolymer of	alcohol dehydrogenase	239
acrylamide and 2-hydroxy-	ribonuclease A	239
ethyl methacrylate	α-D-glucosidase	239
	trypsin	239
	urease	239
CELLULOSE CARBONATE DERIVATIVE		
Cellulose–*trans*-2,3-carbonate	dextranase	156
	β-glucosidase	240, 241
	trypsin	240
WOODWARD'S REAGENTS K AS CONDENSING AGENTS		
CM-cellulose	acetylcholinesterase	153
	deoxyribonuclease	153
	apyrase	153
Copolymer of L-alanine	acetylcholinesterase	153
and L-glutamic acid	deoxyribonuclease	153
	trypsin	242
	α-chymotrypsin	242

(*Continued*)

TABLE 2.4—*Continued (4)*

Carrier	Enzyme	Reference
	apyrase	153
Polyaspartic acid	acetylcholinesterase	153
	deoxyribonuclease	153
	apyrase	153
Polygalacturonic acid	acetylcholinesterase	153
	deoxyribonuclease	153
	apyrase	153
Polymethylmethacrylic acid	apyrase	153
Polypropyrene–*p*-aminostyrene	trypsin	243

CARBODIIMIDE REAGENTS AS CONDENSING AGENTS

AE-cellulose	peroxidase	244
	aminoacylase	80
CM-cellulose	peroxidase	244
CM-Sephadex	pronase	245
Bio-Gel CM-100	β-amylase	246
	pullulanase	246
	trypsin	247
Copolymer of acrylamide	alcohol dehydrogenase	239
and acrylic acid	ribonuclease A	239
	β-D-glucosidase	239
	trypsin	239, 247
	urease	239
Polypropyrene–*p*-aminostyrene	trypsin	243
Glycine derivative of copolymer	trypsin	248
of chloromethylstyrene and		
divinylbenzene		
Aminosilane derivative of	hydrogenase	122
porous glass	acetylcholinesterase	124
	lactase	132
	leucine aminopeptidase	249
	pepsin	250, 251
	papain	141
Succinamidopropyl-glass beads	glutamate dehydrogenase	252

N-ETHOXYCARBONYL–2-ETHOXY–1,2-DIHYDROQUINOLINE AS CONDENSING AGENT

CM-cellulose	trypsin	253
	urease	253
CM-Sephadex	trypsin	253
	urease	253
Poly(4-methacryloxybenzoic	peroxidase	254, 255
acid)	α-amylase	254
	lysozyme	254
	aldolase	254

i) Acid azide derivatives

In the field of peptide synthesis the azide method is a widely used procedure in which no racemization of amino acids is caused. This tech-

nique has been applied for the immobilization of many enzymes.
As shown in Eq. 2.7, CM-cellulose is converted to the methylester, and then to the hydrazide by the action of hydrazine. The hydrazide reacts with nitrous acid to form the corresponding azide derivative (19), which reacts with the enzyme protein at low temperature to give immobilized enzyme.[69]

$$\begin{aligned}
&\text{cellulose–OCH}_2\text{COOH} \xrightarrow{\text{CH}_3\text{OH/HCl}} \text{cellulose–OCH}_2\text{COOCH}_3 \xrightarrow{\text{NH}_2\text{NH}_2} \\
&\text{cellulose–OCH}_2\text{CONHNH}_2 \xrightarrow{\text{NaNO}_2\text{/HCl}} \text{cellulose–OCH}_2\text{CON}_3 \xrightarrow{\text{enzyme}} \\
&\text{cellulose–OCH}_2\text{CONH–enzyme} \qquad\qquad\quad \textbf{19}
\end{aligned} \qquad (2.7)$$

This azide derivative of CM-cellulose is commercially available from Miles-Seravac Laboratories Ltd. Enzacryl AH, a hydrazide derivative of polyacrylamide (20), from Koch-Light Laboratories Ltd. can be converted to its azide derivative and used for the immobilization of enzymes as shown in Eq. 2.8.[83,103]

$$\begin{array}{ccc}
\text{–CHCH}_2\text{–} & \text{–CHCH}_2\text{–} & \text{–CHCH}_2\text{–} \\
| & | & | \\
\text{CONHNH}_2 & \text{CON}_3 & \text{CONH-enzyme} \\
\textbf{20} & &
\end{array}$$

$$-\text{CHCH}_2- \xrightarrow{\text{NaNO}_2\text{/HCl}} -\text{CHCH}_2- \xrightarrow{\text{enzyme}} -\text{CHCH}_2- \qquad (2.8)$$

ii) Carboxychloride resin

Carboxychloride resin (21), prepared by the action of thionylchloride on carboxylic acid resin, can be reacted with enzymes at low temperature for immobilization according to Eq. 2.9.[60,82,105]

$$\underset{\substack{\text{(carboxylic} \\ \text{acid resin)}}}{\text{R–COOH}} \xrightarrow{\text{SOCl}_2} \underset{\textbf{21}}{\text{R–COCl}} \xrightarrow{\text{enzyme}} \text{R–CONH-enzyme} \qquad (2.9)$$

As carboxylic acid resins, Amberlite XE-64, Amberlite IRC-50 and others have been used for the immobilization of enzymes.

In 1956, Brandenberger[60] used a carboxychloride of Amberlite IRC-50 for the immobilization of lipase. This report is considered to be the first of the peptide binding methods.

iii) Maleic anhydride derivative

The copolymers of maleic anhydride with ethylene, styrene, butanedioldivinylether, methyl vinyl, isobutyl vinyl ester and others have been used for the immobilization of many enzymes.[173-180]

For example, as shown in Eq. 2.10, on reacting the copolymer (22) of maleic anhydride and ethylene with enzyme in the presence of hexamethylene diamine, the enzyme is immobilized by the formation of peptide bonds between maleic anhydride and amino groups of the enzyme protein.

$$
\begin{array}{c}
-CH_2-CH-CH-CH_2-CH_2-CH-CH-CH_2-CH_2- \\
| \qquad | \qquad\qquad | \qquad | \\
CO \quad COO^- \qquad CO \quad COO^- \\
| \qquad\qquad\qquad | \\
NH \qquad\qquad\qquad NH \\
| \qquad\qquad\qquad | \\
(CH_2)_6 \qquad\qquad enzyme \\
| \qquad\qquad\qquad | \\
NH \qquad\qquad\qquad NH \\
| \qquad\qquad\qquad | \\
CO \quad COO^- \qquad CO \quad COO^- \\
| \qquad | \qquad\qquad | \qquad | \\
-CH_2-CH-CH-CH_2-CH_2-CH-CH-CH_2-CH_2-
\end{array}
$$

$$
\left[\begin{array}{c} -CH_2-CH_2-CH-CH- \\ | \qquad\; | \\ OC \quad CO \\ \!\!\!\searrow\!\!O\!\!\swarrow \end{array} \right]_n + \text{enzyme} \xrightarrow{\;NH_2(CH_2)_6NH_2\;}
$$

22

$$(2.10)$$

iv) Isocyanate derivative

As shown in Eq. 2.11 or 2.12, by reacting phosgene with a carrier (R–NH$_2$) containing an aromatic amino group at alkaline pH or on warming a carrier (**19**) containing an acid azide group with hydrochloric acid, the corresponding isocyanate derivative (**23**) is obtained. The resulting isocyanate derivatives have been used for the immobilization of enzymes.

$$
R-NH_2 \xrightarrow{COCl_2} \underset{\textbf{23}}{R-NCO} \xrightarrow{enzyme} R-NHCONH-enzyme \qquad (2.11)
$$

$$
\underset{\textbf{19}}{R-CON_3} \xrightarrow{HCl} \underset{\textbf{23}}{R-NCO} \xrightarrow{enzyme} R-NHCONH-enzyme \qquad (2.12)
$$

As water-insoluble carriers for this method, polysaccharides such as CM-cellulose and Sephadex, and polyaminostyrene have been used. For example, the isocyanate derivative (**24**) prepared by the reaction of polyaminostyrene and phosgene was used for immobilization of catalase as shown in Eq. 2.13.

$$
\left[\begin{array}{c} -CHCH_2- \\ \bigcirc \\ | \\ NH_2 \end{array} \right]_n \xrightarrow{COCl_2} \left[\begin{array}{c} -CHCH_2- \\ \bigcirc \\ | \\ NCO \end{array} \right]_n \xrightarrow{enzyme} \left[\begin{array}{c} -CHCH_2- \\ \bigcirc \\ | \\ NHCONH-\text{enzyme} \end{array} \right]_n
$$

24

$$(2.13)$$

v) CNBr-activated polysaccharide

This method was first developed by Axén et al.,[200] and has been used for the immobilization of many enzymes. The method involves the activation of polysaccharide with cyanogen bromide to give an inert carbamate (**26**) and reactive imidocarbonate (**27**) through the intermediate cyano derivative (**25**) according to Eq. 2.14. The reactive imidocarbonate subsequently reacts with enzyme protein as shown in Eq. 2.15.

$$
R\!\!\begin{array}{c}OH\\OH\end{array}\xrightarrow{CNBr}\left[\,R\!\!\begin{array}{c}OC\!\equiv\!N\\OH\\25\end{array}\,\right]\xrightarrow{H_2O}\begin{array}{l}\to R\!\!\begin{array}{c}OCONH_3\\OH\\26\end{array}\\[2mm]\to R\!\!\begin{array}{c}O\\C\!=\!NH\\O\end{array}\\27\end{array}
\qquad (2.14)
$$

$$
R\!\!\begin{array}{c}O\\C\!=\!NH\\O\end{array}+\text{enzyme}\longrightarrow
\left\{
\begin{array}{l}
R\!\!\begin{array}{c}\overset{\displaystyle NH}{\underset{\displaystyle OC\!-\!NH-\text{enzyme}}{\|}}\\OH\\28\end{array}\\[4mm]
R\!\!\begin{array}{c}O\\C\!=\!N-\text{enzyme}\\O\\29\end{array}\\[4mm]
R\!\!\begin{array}{c}\overset{\displaystyle O}{\underset{\displaystyle OC\!-\!NH-\text{enzyme}}{\|}}\\OH\\30\end{array}
\end{array}
\right.
\qquad (2.15)
$$

(with 27 on the left)

Equations (2.14) and (2.15) were based mainly on infrared data indicating the existence of –CO and –C=N– structures and on the finding that, on allowing ethylimidocarbonate to react with amino acids and amino acid derivatives, the main products were indeed *N*-substituted isoureas (28), *N*-substituted imidocarbonates (29), and *N*-substituted carbamates (30).[256] However, additional evidence accumulated in the last few years indicates that major reaction products of cyanogen bromide-activated polysaccharides with amines are probably the substituted isourea structures.[257-259]

This immobilization technique is simple, and binding of enzymes to carriers is performed under mild conditions. Consequently, this is a widely applicable technique and has been applied for the immobilization of many enzymes.

For example, leucine aminopeptidase attached to the isocyanate derivative of fluorochrome was immobilized using CNBr-activated Sepharose.[196] This report on immobilized fluorecent leucine aminopeptidase is of interest in relation to visualizing the enzyme distribution in gel matrices.

vi)   Cellulose carbonate derivative

Barker *et al*.[260] prepared cellulose-*trans*-2,3-carbonate (31) by treating cellulose with ethylchloroformate, and immobilized enzymes by reaction with free amino groups of the enzyme protein. As shown in Eq. 2.16, cellulose powder was suspended in a mixed solvent of methylsulfoxide, dioxane and triethylamine, and ethylchloroformate was added at 0°C. After stirring for 10 min, the reaction mixture was neutralized with conc. HCl and transferred into 90% ethanol to give the reactive cellulose-*trans*-2,3-carbonate (**31**). This activated cellulose was added to an enzyme [dissolved in buffer solution at neutral pH, then the suspension was gently stirred for 4 h at 5°C. By this procedure, the enzyme was easily immobilized according to Eq. 2.17.[156,240,241]

$$R\!\!<\!\!\begin{array}{c} OH \\ OH \end{array} + ClCOOC_2H_5 \longrightarrow R\!\!<\!\!\begin{array}{c} O \\ O \end{array}\!\!>\!\!C=O \qquad (2.16)$$

cellulose   **31**

$$R\!\!<\!\!\begin{array}{c} O \\ O \end{array}\!\!>\!\!C=O + \text{enzyme} \longrightarrow \begin{cases} R\!\!<\!\!\begin{array}{c} O \\ O \end{array}\!\!>\!\!C=N\text{–enzyme} \\[2ex] R\!\!<\!\!\begin{array}{c} OCONH\text{–enzyme} \\ OH \end{array} \end{cases} \qquad (2.17)$$

31

    This method is simple, and attachment of the enzyme to the carrier can be performed under very mild conditions, as in the case of the CNBr-activated polysaccharide method. In addition, immobilized preparations are produced with various modes of binding.

vii) Condensing reagents
    As condensing reagents, carbodiimide reagents such as dicyclohexyl-carbodiimide (**32**), 1-ethyl-3-(3-dimethylaminopropyl) carbodiimide hydrochloride (**33**) and 1-cyclohexyl-3-(2-morpholinoethyl) carbodiimide metho-*p*-toluene sulfonate (**34**), Woodward's reagent K (*N*-ethyl-5-phenyl-isoxazolium-3'-sulfonate) (**35**), and *N*-ethoxycarbonyl-2-ethoxy-1,2-dihydroquinoline (**36**) are commonly used, and these reagents are now commercially available.
1) Binding to amino groups of the enzyme
    As shown in Eq. 2.18 carbodiimides react with carriers containing carboxyl groups to give the *O*-acylisourea derivative, and the activated carboxyl groups subsequently react with amino groups of the enzyme to produce immobilized enzyme.

$$\langle H \rangle\text{-}N=C=N\text{-}\langle H \rangle \qquad CH_3CH_2\text{-}N=C=N\text{-}CH_2CH_2CH_2\text{-}^+N\!\!\begin{array}{c}CH_3\\ CH_3\end{array}\cdot\ HCl$$

$$\textbf{32} \qquad\qquad\qquad\qquad \textbf{33}$$

$$\langle H \rangle\text{-}N=C=N\text{-}CH_2CH_2\text{-}^+N\!\!\bigcirc\!\!O \cdot {}^-O_3S\text{-}\langle O \rangle\text{-}CH_3$$

$$\textbf{34}$$

$$\textbf{35} \qquad\qquad\qquad\qquad \textbf{36}$$

$$R\text{-}COOH + \begin{array}{c}R'\\ |\\ N\\ \|\\ C\\ \|\\ N\\ |\\ R''\end{array} \longrightarrow R\text{-}COO\text{-}\begin{array}{c}R'\\ |\\ NH\\ |\\ C\\ \|\\ N\\ |\\ R''\end{array} \xrightarrow{H_2N\text{-enzyme}} R\text{-}CONH\text{-enzyme} + R'\text{-}NH\text{-}\overset{\displaystyle O}{\overset{\displaystyle \|}{C}}\text{-}NH\text{-}R''$$

$$(2.18)$$

As shown in Eqs. 2.19 and 2.20, Woodward's reagent K and N-ethoxycarbonyl-2-ethoxy-1,2-dihydroquinoline can be used in the same way as in Eq. 2.18.

As carriers containing carboxyl groups, CM-cellulose, CM-Sephadex, Bio-Gel CM-100, polyaspartic acid, polygalacturonic acid, polymethylmethacrylate and succinized porous glass have been used.

2) Binding to carboxyl groups of enzymes

A carrier containing amino groups and a carbodiimide reagent or Woodward's reagent K are added to enzyme solution and stirred. Peptide bonds are formed between amino groups of the carrier and carboxyl groups of the enzyme protein, immobilizing the enzyme.

As carriers containing amino groups, AE-cellulose, amino-silanized porous glass and others have been used.

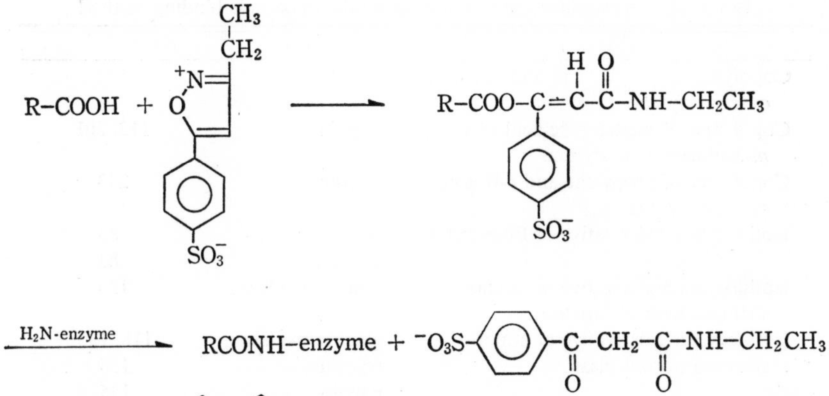

$$R-COOH + \underset{\underset{SO_3^-}{\overset{CH_3}{\overset{|}{CH_2}}}}{\overset{+N=}{O}} \longrightarrow R-COO-C=C-C-NH-CH_2CH_3$$

(with $\overset{H}{\overset{|}{}} \overset{O}{\overset{\|}{}}$ on the product and SO_3^- on the ring)

$$\xrightarrow{\text{H}_2\text{N-enzyme}} RCONH\text{-enzyme} + {}^-O_3S-\langle \rangle-C-CH_2-C-NH-CH_2CH_3$$

(with $\underset{O}{\overset{\|}{}} \quad \underset{O}{\overset{\|}{}}$)

$$R-COOH + \underset{\underset{COOC_2H_5}{N}}{\bigcirc\bigcirc}OC_2H_5 \longrightarrow \qquad (2.19)$$

$$R-\overset{O}{\overset{\|}{C}}-O-\overset{O}{\overset{\|}{C}}-O-C_2H_5 + \bigcirc\bigcirc_N + C_2H_5OH$$

$$\bigg\downarrow \text{H}_2\text{N-enzyme}$$

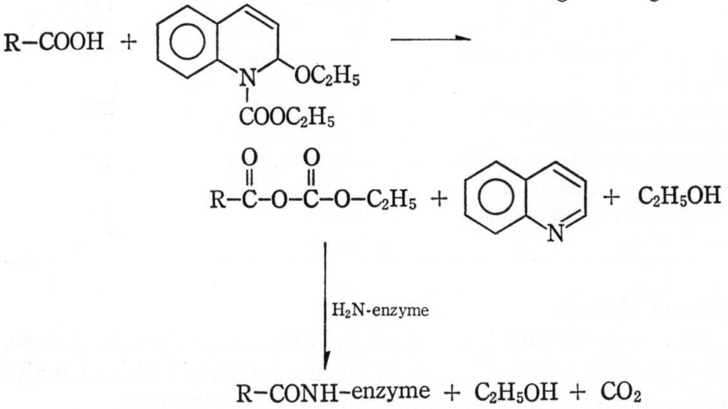

(2.20)

$$R\text{-CONH-enzyme} + C_2H_5OH + CO_2$$

viii) Thioamide binding

The principle of the thioamide binding method is the same as that of the method using isocyanate derivatives, and it is based on the formation of thioamide bonds between free amino groups of the enzyme protein and isothiocyanate derivatives of carriers. Namely, as shown in Eq. 2.21, a carrier ($R-NH_2$) containing an amino group is reacted with thiophosgene to give the corresponding isothiocyanate derivative (**37**), and an enzyme is immobilized by reaction with the derivative.

$$R-NH_2 \longrightarrow \underset{\mathbf{37}}{R-NCS} \xrightarrow{\text{enzyme}} R-NHCSNH\text{-enzyme} \qquad (2.21)$$

As carriers, copolymers of acrylic acid-*m*-aminostyrene and methacrylic acid-*m*-aminostyrene, as well as aminosilanized porous glass, hydroxylapatite, nickel-alumina and silica have been used.

Immobilized enzymes which have been prepared by this method are listed in Table 2.5.

TABLE 2.5   Immobilization of enzymes by the thioamide binding method

Carrier	Enzyme	Reference
Copolymer of acrylic acid and *m*-isothiocyanate styrene	papain	112
Copolymer of methacrylic acid and *m*-isothiocyanate styrene	papain	112, 261
Copolymer of propylene and *p*-isothiocyanate styrene	trypsin	243
Isothiocyanate derivative of Enzacryl AA	α-amylase	83
	β-amylase	83
Isothiocyanate derivative of aminosilanized hydroxylapatite	glucose oxidase	135
Isothiocyanate derivative of aminosilanized porous glass	glucose oxidase	118, 135
	trypsin	135
	papain	135
	ficin	135
Isothiocyanate derivative of aminosilanized alumina	glucose oxidase	135
Isothiocyanate derivative of aminosilanized nickel-alumina	glucose oxidase	262
Isothiocyanate derivative of aminosilanized nickel-nickel oxide	glucose oxidase	263
Isothiocyanate derivative of aminosilanized silica	trypsin	135
	papain	135

c. Alkylation Method

The alkylation method is based on the alkylation of amino groups, phenolic groups or sulfhydryl groups of an enzyme protein with a reactive group such as halide in water-insoluble carriers (Table 2.6).

TABLE 2.6   Immobilization of enzymes by the alkylation method

Carrier	Enzyme	Reference
HALOGENOACETYL DERIVATIVES		
Chloroacetyl cellulose	aminoacylase	264
Bromoacetyl cellulose	ribonuclease	265
	glucoamylase	266
	trypsin	265
	α-chymotrypsin	265
	pronase	245
	aminoacylase	264
Iodoacetyl cellulose	glucoamylase	266
	aminoacylase	264
Polyethyleneglycol iodoacetyl cellulose	urease	89
TRIAZINYL (CYANUR) DERIVATIVES		
Dichloro–*s*-triazinyl cellulose	alcohol dehydrogenase	267

(Continued)

Immobilization of enzymes 35

TABLE 2.6—*Continued*

Carrier	Enzyme	Reference
	lactate dehydrogenase	267–269
	tyrosinase	270
	catalase	38
	creatine kinase	268
	pyruvate kinase	268
	ribonuclease	38
	phosphodiesterase	271
	phosphomonoesterase	271
	glucoamylase	272, 273
	lactase	268, 274
	invertase	16
	trypsin	38
	α-chymotrypsin	38, 275, 276
	penicillin amidase	277, 278
	ATPase	166
	apyrase	166
Dichloro–s-triazinyl Sephadex	α-chymotrypsin	276
Dichloro–s-triazinyl Sepharose	hyaluronidase	279
	α-chymotrypsin	276
Dichloro–s-triazinyl porous glass	phosphorylase	33
Dichloro–s-triazinyl bentonite	invertase	18
OTHER HALOGENO DERIVATIVES		
Copolymer of methacrylic acid and *m*-fluorodinitro anilide	alcohol dehydrogenase	280, 281
	alkaline phosphatase	171
	invertase	281, 282
	diastase	280, 282
	pepsin	280, 281
	aminoacylase	78, 79
	L-glutamate dehydrase	82
Copolymer of methacrylic acid and 4-fluorostyrene	diastase	283
	papain	261, 283
Copolymer of methacrylic acid and methacrylate 4-iodo–*n*-butylester	trypsin	284
	α-chymotrypsin	285
	δ-chymotrypsin	285
	urease	284
Polymethacrylate–4-iodo–*n*-butylester	urease	286

Halogenoacetyl, triazinyl or halogenomethacryl derivatives have been used as carriers for this method. The halogenoacetyl derivatives include chloroacetyl cellulose, bromoacetyl cellulose, iodoacetyl cellulose, polyethyleneglycoliodoacetyl cellulose and so on. For example, as shown in

cellulose-OH $\xrightarrow[\text{BrCH}_2\text{COOH/dioxane}]{\text{BrCH}_2\text{COBr/}}$ cellulose-OCOCH$_2$Br $\xrightarrow{\text{enzyme}}$
cellulose-OCOCH$_2$-enzyme   (2.22)

Eq. 2.22, trypsin, chymotrypsin and ribonuclease were immobilized by using bromoacetyl cellulose (38) prepared by the reaction of cellulose with bromoacetylbromide in dioxane-bromoacetic acid.[265]

We[264] have also extensively investigated the immobilization of mold aminoacylase by using bromoacetyl cellulose, and found that when the enzyme was reacted with bromoacetyl cellulose in the presence of a salting-out agent such as ammonium sulfate, active and stable immobilized enzyme was obtained. Further, we prepared iodoacetyl cellulose (39) by treating bromoacetyl cellulose with sodium iodide in alcohol as shown in Eq. 2.23.

$$\text{cellulose-OCOCH}_2\text{Br} \xrightarrow{\text{NaI/alcohol}} \text{cellulose-OCOCH}_2\text{I} \xrightarrow{\text{enzyme}}$$
$$\underset{38}{} \qquad\qquad \underset{39}{}$$

cellulose-OCOCH$_2$-enzyme   (2.23)

The activity of the immobilized aminoacylase prepared with iodoacetyl cellulose was higher than that of enzyme immobilized using bromoacetyl cellulose.

Katchalski et al.[89] carried out immobilization of urease by the alkylation method as follows. As shown in Eq. 2.24, a terminal hydroxyl group of polyethyleneglycol was esterified with monoiodoacetic acid and another terminal hydroxyl group was reacted with phosgene to give the chloroformate. This derivative was reacted with cellulose to give an iodoacetyl cellulose derivative of polyethyleneglycol (40). Using this reactive carrier, urease was immobilized; this was the first report on immobilization by the alkylation method.

(2.24)

As carriers possessing reactive halogens, triazinyl derivatives of cellulose, Sephadex, Sepharose, porous glass and bentonite have been employed for the immobilization of many enzymes. For example, as shown in Eq. 2.25, dichloro-*s*-triazinyl-cellulose (cyanul cellulose) (**41**) was

$$\text{cellulose} + \text{Cl} - \underset{\underset{\displaystyle N=C}{\overset{\displaystyle N-C}{|}}}{C} \quad \overset{\text{Cl}}{\underset{\text{Cl}}{N}} \longrightarrow \text{cellulose} - O - \underset{\underset{\displaystyle N=C}{\overset{\displaystyle N-C}{|}}}{C} \quad \overset{\text{Cl}}{\underset{\text{Cl}}{N}}$$

41

$$\xrightarrow{\text{enzyme}} \text{cellulose} - O - \underset{\underset{\displaystyle N=C}{\overset{\displaystyle N-C}{|}}}{C} \quad \overset{\text{Cl}}{\underset{\text{enzyme}}{N}} \qquad (2.25)$$

prepared by the reaction of cellulose and trichlorotriazine at pH 9–11, and this active halogeno compound has been reacted with chymotrypsin and many other enzymes to give immobilized enzymes.

Halogen derivatives of polymethacrylate have also been used in this method.

Namely, using copolymers of methacrylate and 4-fluorostyrene (**42**) or of methacrylate and methacrylate-*m*-fluorodinitroanilide (**43**), invertase,

$$(2.26)$$

$$(2.27)$$

diastase and other enzymes were immobilized according to Eqs. 2.26 and 2.27.[280,281,283]

Further, as shown in Eqs. 2.28 and 2.29, urease, chymotrypsin, trypsin and other enzymes have been immobilized by using the copolymer of methacrylate and methacrylate-4-iodo-n-butylester (**44**) or the homopolymer of methacrylate-4-iodo-n-butylester (**45**).[284-286]

$$
\begin{bmatrix}
 & CH_3 & & CH_3 \\
-CH_2-\overset{|}{\underset{|}{C}}-CH_2-\overset{|}{\underset{|}{C}}- \\
& CO & & COOH \\
& O \\
& (CH_2)_4I
\end{bmatrix}_n
+ \ \text{enzyme} \longrightarrow
\begin{bmatrix}
 & CH_3 & & CH_3 \\
-CH_2-\overset{|}{\underset{|}{C}}-CH_2-\overset{|}{\underset{|}{C}}- \\
& CO & & COOH \\
& O \\
& (CH_2)_4\text{- enzyme}
\end{bmatrix}_n
$$

44

$$(2.28)$$

$$
\begin{bmatrix}
& CH_3 \\
-CH_2-\overset{|}{\underset{|}{C}}-CH_2- \\
& CO \\
& (CH_2)_4I
\end{bmatrix}_n
+ \ \text{enzyme} \longrightarrow
\begin{bmatrix}
& CH_3 \\
-CH_2-\overset{|}{\underset{|}{C}}-CH_2- \\
& CO \\
& (CH_2)_4\text{-enzyme}
\end{bmatrix}_n
$$

45

$$(2.29)$$

d. Carrier Binding with Bifunctional Reagents

This method is based on the formation of cross-links between the amino groups of carriers and the amino groups of enzyme protein by means of bi- or multifunctional reagents. For convenience, we will refer to this method as the "carrier cross-linking method."

Glutaraldehyde is most commonly employed as the functional reagent, and many enzymes have been immobilized by the formation of Schiff bases between the amino groups of carriers and of enzyme protein. Enzymes which have been immobilized on carriers by means of glutaraldehyde are listed in Table 2.7.

As carriers possessing amino groups, aminoethyl cellulose (AE-cellulose), DEAE-cellulose, amino derivatives of Sepharose, inactivated albumin, partially deacylated chitin, heated microbial cells, aminoethyl polyacrylamide (AE-polyacrylamide), aminosilane derivatives of porous glass, etc., have been used. For example, enzymes such as aldolase, glyceroaldehydephosphate dehydrogenase and trypsin have been immobilized by using AE-cellulose and glutaraldehyde as shown in Eq. 2.30.[33,81]

TABLE 2.7 Immobilization of enzymes on carriers with glutaraldehyde

Carrier	Enzyme	Reference
AE-cellulose	glyceraldehyde phosphate dehydrogenase	81
	lactase	287
	pepsin	288
	rennin	227
	trypsin	289, 290
	aldolase	81
	fructosediphosphatase	81
DEAE-cellulose	catalase	291
Amino derivative of Sepharose	lactate dehydrogenase	192
	malate dehydrogenase	192
	citrate synthase	192
Partially deacylated chitin	acid phosphatase	292
	lactase	292
	trypsin	292
	α-chymotrypsin	292
	glucose isomerase	293
Inactivated albumin particles	alcohol dehydrogenase	294
	lactate dehydrogenase	294
	glucose oxidase	294
	L-amino acid oxidase	294
	uricase	294
	catalase	294
	asparaginase	294
	urease	294
	carbonic anhydrase	294
	phenylalanine decarboxylase	294
Heated microbial cells	β-amylase	295
	glucoamylase	295
	invertase	295
	trypsin	295
	glucose isomerase	295
AE–polyacrylamide	glucose oxidase	296
	ribonuclease	296
	acid phosphatase	296
	trypsin	296
	α-chymotrypsin	296
Polypropylene–p-aminostyrene	trypsin	243
Aminoalkylated porous glass	lactate dehydrogenase	114, 297, 298
	uricase	121, 298
	hydrogenase	122
	hepatic microsomal flavoprotein oxidase	213
	phosphorylase	33, 216
	carbamate kinase	299
	polynucleotide phosphorylase	189

(*Continued*)

TABLE 2.7—*Continued*

Carrier	Enzyme	Reference
	tannase	300
	glucoamylase	127, 301
	lactase	132, 302, 303
	trypsin	138, 297, 304–306
	neutral protease	143
	apyrase	305
	ribulose diphosphate carboxylase	145
Aminoarylated porous glass	lactate dehydrogenase	116
	pepsin	307, 308
	α-chymotrypsin	307
Aminoalkylated zirconium oxide porous glass	papain	141
Aminoalkylated alumina	alcohol dehydrogenase	309
Aminoalkylated ceramic	aminoacylase	310
Aminoalkylated Englehard material	glucose oxidase	311
	catalase	311
Aminoalkylated Harshaw material	glucose oxidase	311
	catalase	311
Aminoalkylated Hornbende material	glucoamylase	312
	invertase	313
Aminoalkylated magnetite	invertase	314
	trypsin	314
Aminoalkylated nickel-silica-alumina	glucose oxidase	315, 316
	catalase	315, 316
Aminoalkylated Norton material	glucose oxidase	311
	catalase	311
Aminoalkylated sand	alcohol dehydrogenase	317
	lactate dehydrogenase	317
	urease	317
Aminoalkylated silica	glucoamylase	318

$$\begin{aligned} &\text{cellulose–OCH}_2\text{CH}_2\text{NH}_2 + \text{OHC(CH}_2)_3\text{CHO} + \text{H}_2\text{N–enzyme} \\ &\longrightarrow \text{cellulose–OCH}_2\text{CH}_2\text{N}=\text{CH(CH}_2)_3\text{CH}=\text{N–enzyme} \end{aligned} \qquad (2.30)$$

Hexamethylenediisocyanate has also been used as a bifunctional reagent for immobilization, e.g., in the case of aminoacylase, as shown in Eq. 2.31.[80]

$$\begin{aligned} &\text{cellulose–OCH}_2\text{CH}_2\text{NH}_2 + \text{OCN(CH}_2)_6\text{NCO} + \text{H}_2\text{N–enzyme} \\ &\longrightarrow \text{cellulose–OCH}_2\text{CH}_2\text{NHCONH(CH}_2)_6\text{NHCONH–enzyme} \end{aligned} \qquad (2.31)$$

It is also possible to treat with glutaraldehyde enzymes which are physically adsorbed on carriers containing no functional groups. For example, enzymes were physically adsorbed on a core such as glass beads, carbon, silica, alumina, titania, etc., or ionically bound on silica treated with polyethyleneimine, and then the adsorbed proteins were treated with glutaraldehyde to give immobilized preparations.[33,319-324]

e. Carrier Binding by the Ugi Reaction

Recently, reactions involving highly reactive isocyanide compounds have been applied for the immobilization of enzymes.

Ugi *et al.*[325,326] reported that the reactions of carboxylate, amine, aldehyde or ketone, and isocyanide, lead to the formation of an *N*-substituted amide according to Eq. 2.32.

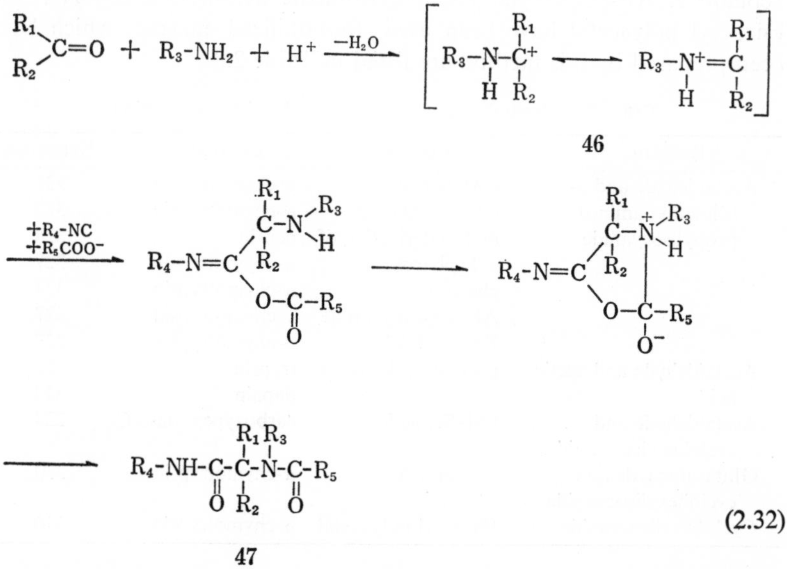

$$
\begin{array}{c}
R_1 \\
\\
R_2
\end{array}\!\!\!\!C{=}O \;+\; R_3{-}NH_2 \;+\; H^+ \xrightarrow{-H_2O}\;
\left[\begin{array}{cc}
 & R_1 \\
R_3{-}N{-}C^+ & \longleftrightarrow \quad R_3{-}N^+{=}C \\
\;H\;\;R_2 & \;\;H\;\;R_2
\end{array}\right]
$$

46

$$\xrightarrow[+R_5COO^-]{+R_4{-}NC}\quad R_4{-}N{=}C \overset{\displaystyle C{-}N\langle^{R_3}_{H}}{\underset{O{-}C{-}R_5}{}} \longrightarrow\; R_4{-}N{=}C$$

$$\longrightarrow\quad R_4{-}NH{-}\underset{O}{\overset{}{C}}{-}\underset{R_2}{\overset{R_1}{C}}{-}\underset{}{\overset{R_3}{N}}{-}\underset{O}{\overset{}{C}}{-}R_5$$

(2.32)

47

That is, an amino compound (R_3) reacts with a carbonyl compound (R_1, R_2) to produce an immonium ion (**46**), and this ion subsequently couples with isocyanide and carboxyl compounds (R_4, R_5) to lead to the formation of an amide (**47**). Consequently, if one of $R_1{\sim}R_5$ is a water-insoluble carrier, immobilization of enzymes can be carried out. For instance, as

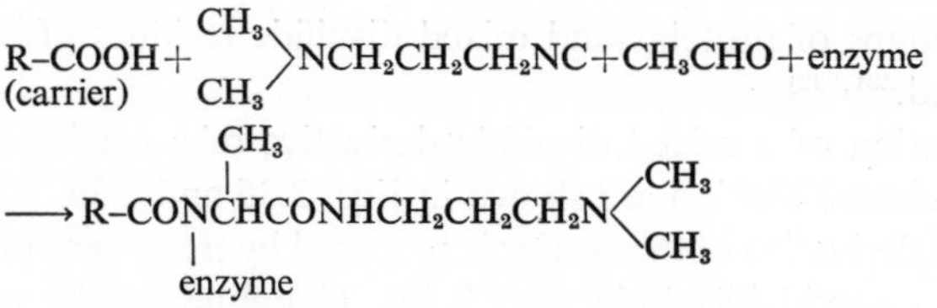

$$R{-}COOH + \overset{CH_3\searrow}{\underset{CH_3\nearrow}{}}NCH_2CH_2CH_2NC + CH_3CHO + enzyme$$
(carrier)

$$\longrightarrow R{-}CON\overset{CH_3}{\underset{enzyme}{CH}}CONHCH_2CH_2CH_2N\langle^{CH_3}_{CH_3}$$
(2.33)

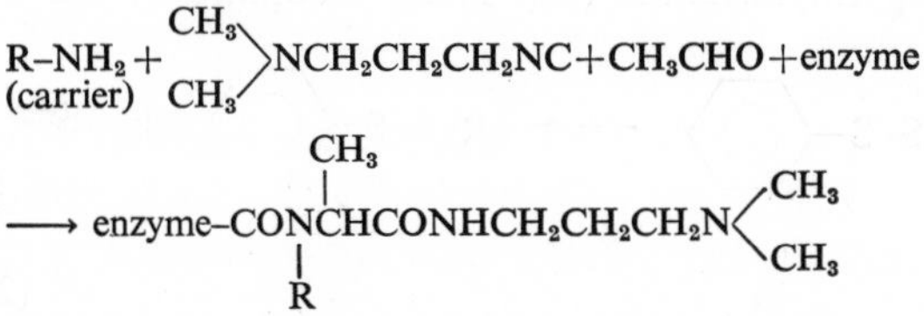

$$R{-}NH_2 + \overset{CH_3\searrow}{\underset{CH_3\nearrow}{}}NCH_2CH_2CH_2NC + CH_3CHO + enzyme$$
(carrier)

$$\longrightarrow enzyme{-}CON\overset{CH_3}{\underset{R}{CH}}CONHCH_2CH_2CH_2N\langle^{CH_3}_{CH_3}$$
(2.34)

shown in Eqs. 2.33 and 2.34, a water-insoluble carrier containing amino or carboxyl groups was suspended in enzyme solution, and to this suspension acetoaldehyde and 3-(dimethylamino) propylisocyanide were added. The reaction mixture was stirred for 6 h at pH 6.5 with 0.5N HCl to prepare the immobilized enzyme.[327]

As carriers, CM-Sephadex, CM-Sepharose, amino derivatives of Sepharose, AE-polyacrylamide, polyisonitrile nylon, Enzacryl AA and Enzacryl polyacetal have been used. Immobilized enzymes which have been prepared by this method are listed in Table 2.8.

TABLE 2.8   Immobilization of enzymes by the Ugi reaction

Reagent	Carrier	Enzyme	Reference
Acetaldehyde and 3-	CM–Sephadex	α-chymotrypsin	327
(dimethyl-amino)	CM–Sepharose	α-chymotrypsin	327
propylisocyanide	amino derivative of	pepsin	328
	Sepharose	α-chymotrypsin	327
	chelatin	α-chymotrypsin	327
	AE-polyacrylamide	α-chymotrypsin	327
	Enzacryl AA	α-chymotrypsin	327
Acetaldehyde and acetic	polyisonitrile nylon	trypsin	329
acid		papain	329
Acetaldehyde and	SM–Sephadex	carboxypeptidase C	224
cyclohexylisocyanide			
Glutaraldehyde and	Enzacryl AA	α-chymotrypsin	330
cyclohexylisocyanide			
Cyclohexylisocyanide	Enzacryl polyacetal	α-chymotrypsin	330

In this method, enzymes can be coupled through either their amino groups with carboxyl groups in the carriers or their carboxyl groups with amino groups in the carriers.

f. Carrier Binding by Thiol-disulfide Interchange

Recently, an immobilization method by thioldisulfide interchange between thiol groups of proteins and mixed disulfide residues of carriers has been reported.[331,332]

The preparation of a mixed disulfide derivative of a carrier and its use for enzyme immobilization are shown in Eqs. 2.35 and 2.36. Namely a mixed disulfide derivative of the carrier is prepared by treating a thiolated carrier with 2,2′-dipyridyldisulfide (Eq. 2.35). Coupling of the enzyme

$$(2.35)$$

$$R-S-S-\text{\scriptsize$\langle$}\bigcirc\text{\scriptsize$\rangle$}_{N} + HS-enzyme \longrightarrow R-S-S-enzyme + S=\text{\scriptsize$\langle$}\bigcirc\text{\scriptsize$\rangle$}_{N-H}$$

$$(2.36)$$

through its thiol groups can be accomplished with the liberation of 2-thiopyridone (Eq. 2.36).

As mixed disulfide derivatives of carriers, agarose-glutathione-2-pyridyl disulfide **(48)**, agarose-mercaptohydroxypropylether-2-pyridyl disulfide **(49)** and agarose-adipic acid hydrazide-*N*-acetyl-homocysteine-2-pyridyl disulfide **(50)** have been used. Of these carriers, agarose-glutathione-2-pyridyl disulfide is commercially available under the name of activated-thiol Sepharose from Pharmacia Fine Chemicals.

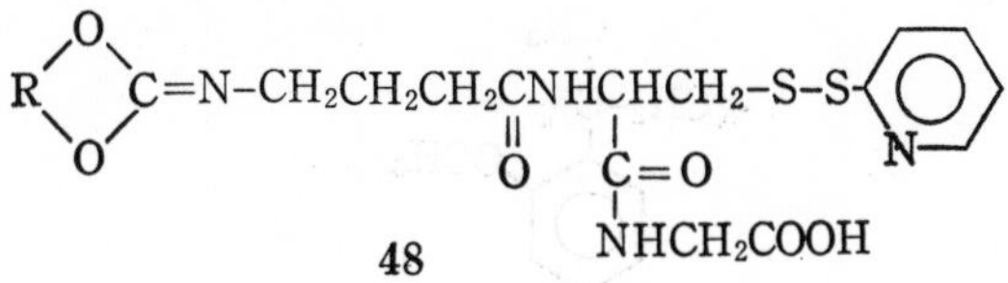

$$R\underset{O}{\overset{O}{\diagup}}C=N-CH_2CH_2CH_2CNHCHCH_2-S-S-\text{\scriptsize$\langle$}\bigcirc\text{\scriptsize$\rangle$}_{N}$$
$$\underset{O}{\overset{\|}{}} \quad \underset{NHCH_2COOH}{\overset{C=O}{|}}$$

48

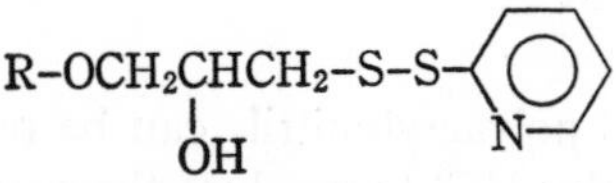

$$R-OCH_2CHCH_2-S-S-\text{\scriptsize$\langle$}\bigcirc\text{\scriptsize$\rangle$}_{N}$$
$$\underset{OH}{|}$$

49

$$R-NHNHCO(CH_2)_4CONHNHCOCHCH_2CH_2-S-S-\text{\scriptsize$\langle$}\bigcirc\text{\scriptsize$\rangle$}_{N}$$
$$\overset{NHCOCH_3}{\underset{|}{}}$$

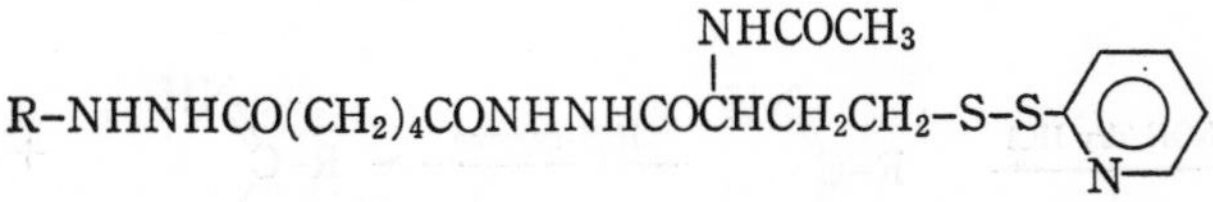

50

Reports on the immobilization of enzymes by this method have been limited to the immobilization of urease.

On the other hand, thiolation of enzymes containing no thiol group, such as chymotrypsin or α-amylase, and subsequent immobilization via disulfide linkages has been also studied.[333] Thiolation of the enzyme was performed using methyl-3-mercaptopropioimidate under slightly alkaline conditions according to Eq. 2.37. The thiolated enzyme was then reacted with the activated carrier to give an immobilized preparation by disulfide linkage.

$$HSCH_2CH_2-C{\overset{\displaystyle +NH_2}{\underset{\displaystyle OCH_3}{\big<}}} + H_2N\text{-emzyme} \longrightarrow$$

$$\text{enzyme-NH-}\overset{\displaystyle +NH_2}{\overset{\displaystyle \|}{C}}CH_2CH_2SH + CH_3OH \qquad (2.37)$$

g. Miscellaneous Carrier Binding Methods

In addition to the above methods, enzymes have been immobilized by other covalent binding methods.

For example, immobilization of enzymes has been performed by using dialdehyde starch, prepared by periodic acid oxidation of starch in the case of papain[334] or by using poly(4-formyl-methoxyphenol)methacrylate (51) for trypsin,[335] invertase[336] and glucose oxidase[336] via Schiff-base linkage between the enzyme amino groups and aldehyde groups of the carriers.

$$\begin{array}{c} CH_3 \\ | \\ -\!\!\!\;(CH_2-C)_{\!n} \\ \end{array}$$

51

Further, as shown in Eq. 2.38 polyacrylonitrile can be reacted with absolute ethanol with bubbling of dry HCl to produce the corresponding imidoester derivative. Using this imidoester, urease, glucoamylase and lactate dehydrogenase have been immobilized.[338]

$$R\text{-CN} \xrightarrow{C_2H_5OH/dry\ HCl} R-C{\overset{\displaystyle \nearrow NH}{\underset{\displaystyle OC_2H_5}{\big\backslash}}} \xrightarrow{H_2N\text{-enzyme}} R-C{\overset{\displaystyle \nearrow NH}{\underset{\displaystyle NH\text{-enzyme}}{\big\backslash}}} + C_2H_5OH$$

$$(2.38)$$

Moreover, as shown in Eq. 2.39, carriers containing amino groups can be activated with CNBr and reacted with amino groups of enzymes for immobilization by the formation of guanidine linkages between amino

$$R\text{-NH}_2 \xrightarrow{CNBr} R\text{-NH-C}\!\equiv\!N \xrightarrow{H_2N\text{-enzyme}} R\text{-NH-}\overset{\displaystyle C}{\underset{\displaystyle \|}{}}\text{-NH-enzyme}$$
$$NH$$

$$(2.39)$$

groups of the enzymes and carriers. By this method, α-chymotrypsin, trypsin and subtilisin have been immobilized.[339]

Another interesting immobilization method for enzymes is the immobilization of tryptophanase, which requires pyridoxal phosphate (PALP) as a coenzyme, by linking the apoenzyme to the immobilized cofactor on Sepharose.[340] Namely, as shown in Eq. 2.40, PALP was

(2.40)

reacted with diazotized *p*-aminobenzamidehexyl-Sepharose (**52**) to give the PALP-Sepharose derivative. Apotryptophanase was then attached by the formation of a Schiff-base linkage between the 4-formyl group of the bound PALP and the ε-amino group of a lysine residue at the active center of the enzyme. The Schiff base was then stabilized by reduction with NaBH$_4$. The reaction procedure is shown in Fig. 2.2.

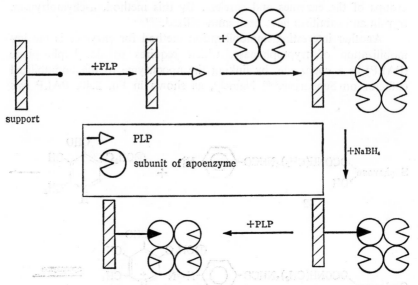

Fig. 2.2   Immobilized tryptophanase.[340]

2.1.2   Cross-linking Method

This immobilization method is based on the formation of chemical bonds, as in the covalent binding method, but water-insoluble carriers are not used in this method. The immobilization of enzymes is performed by the formation of intermolecular cross-linkages between the enzyme molecules by means of bi- or multifunctional reagents.

As cross-linking reagents, glutaraldehyde (Schiff base), isocyanate derivatives (peptide bond), bisdiazobenzidine (diazo coupling), N,N'-polymethylene bisiodoacetoamide (alkylation) and N,N'-ethylene bismaleimide (peptide bond) have been employed.

The functional groups of enzyme proteins participating in the reactions include the α-amino group at the amino terminus, the ε-amino group of lysine, the phenolic group of tyrosine, the sulfhydryl group of cysteine and the imidazole group of histidine.

The cross-linking reactions are carried out under relatively severe conditions, as in the case of covalent binding methods. Thus, in some cases, the conformation of active center of the enzyme may be affected by the reaction, leading to significant loss of activity.

Immobilized enzymes which have been prepared by this method are listed in Table 2.9.

TABLE 2.9   Immobilization of enzymes by the cross-linking method

Cross-linking agent	Enzyme	Reference
Bisdiazobenzidine	papain	96
N,N'-Ethylene bismaleimide	α-amylase	341
Hexamethylene diisocyanate	ribonuclease	342, 343
	α-amylase	343
	trypsin	343
	α-chymotrypsin	343
Toluene diisocyanate	aminoacylase	80
Hexamethylene diisothiocyanate	α-chymotrypsin	342
N,N'-Polymethylene bisiodo-acetamide	aldolase	341, 344, 345
Glutaraldehyde	alcohol dehydrogenase	346
	glutamate dehydrogenase	347
	glucose oxidase	348, 349
	tyrosinase	348
	catalase	350, 351
	peroxidase	348, 349
	ribonuclease	348
	carboxypeptidase A	352–354
	alkaline phosphatase	348
	lysozyme	349
	trypsin	349, 355
	α-chymotrypsin	356, 357
	papain	358–360
	subtilisin	361
	aminoacylase	80

Among cross-linking reagents, glutaraldehyde has been most commonly used for the immobilization of enzymes.

For example, in 1964, Quiocho and Richards[352] treated carboxypeptidase A with glutaraldehyde and obtained the immobilized enzyme by intermolecular cross-linkage, as shown in Eq. 2.41. This may be the oldest report on this cross-linking method.

$$
\begin{array}{c}
-CH=N-\text{enzyme}-N=CH(CH_2)_3CH=N-\text{enzyme}-N=CH- \\
\mid \\
N \\
\parallel \\
CH \\
\mid \\
OHC(CH_2)_3CHO + \text{enzyme} \longrightarrow \quad (CH_2)_3 \\
\mid \\
CH \\
\parallel \\
N \\
\mid \\
-CH=N-\text{enzyme}-N=CH-
\end{array}
\qquad (2.41)
$$

α-Amylase, α-chymotrypsin and ribonuclease were also immobilized according to Eq. 2.42 or 2.43 by using hexamethylene diisocyanate or hexamethylene diisothiocyanate.[342,343]

$OCN(CH_2)_6NCO+$ enzyme ──▸

$$-NHCONH\text{-enzyme-}NHCONH(CH_2)_6NHCONH\text{-enzyme-}NHCONH-$$
$$\overset{|}{NH}$$
$$\overset{|}{CO}$$
$$\overset{|}{NH}$$
$$\overset{|}{(CH_2)_6}$$
$$\overset{|}{NH}$$
$$\overset{|}{CO}$$
$$\overset{|}{NH}$$
$$-NHCONH\text{-enzyme-}NHCONH(CH_2)_6- \qquad (2.42)$$

$SCN(CH_2)_6NCS+$ enzyme ──▸

$$-NHCSNH\text{-enzyme-}NHCSNH(CH_2)_6NHCSNH\text{-enzyme-}NHCSNH-$$
$$\overset{|}{NH}$$
$$\overset{|}{CS}$$
$$\overset{|}{NH}$$
$$\overset{|}{(CH_2)_6}$$
$$\overset{|}{NH}$$
$$\overset{|}{CS}$$
$$\overset{|}{NH}$$
$$-NHCSNH\text{-enzyme-}NHCSNH(CH_2)_6- \qquad (2.43)$$

Further, N,N'-polymethylenebisiodoacetoamide or N,N'-ethylene-bismaleimide has been employed for the immobilization of aldolase in rabbit muscle[341,344,345] or α-amylase,[341] as shown in Eq. 2.44 and 2.45.

$ICH_2CONH(CH_2)_nNHCOCH_2I$ + enzyme ──▸

$$-NHCOCH_2\text{-enzyme-}CH_2CONH(CH_2)_nNHCOCH_2\text{-enzyme-}CH_2CONH-$$
$$\overset{|}{CH_2}$$
$$\overset{|}{CO}$$
$$\overset{|}{NH}$$
$$\overset{|}{(CH_2)_n}$$
$$\overset{|}{NH}$$
$$\overset{|}{CO}$$
$$\overset{|}{CH_2}$$
$$-NHCOCH_2\text{-enzyme-}CH_2CONH(CH_2)_nNH- \qquad (2.44)$$

$$CH-C\overset{O}{\diagdown}\underset{CH-C\diagup}{\overset{\|}{}}N-CH_2-CH_2-N\overset{\overset{O}{\diagdown}C-CH}{\underset{\diagup C-CH}{\|}} + enzyme\longrightarrow$$

$$-CH=CHCO-enzyme-COCH=CHCONH(CH_2)_2NHCOCH=CHCO-enzyme-COCH=CH-$$

$$\begin{array}{c}
| \\
CO \\
| \\
CH \\
\| \\
CH \\
| \\
CO \\
| \\
NH \\
| \\
(CH_2)_2 \\
| \\
NH \\
| \\
CO \\
| \\
CH \\
\| \\
CH \\
| \\
CO \\
|
\end{array}$$

$$-CH=CHCO-enzyme-COCH=CHCONH(CH_2)_2NHCO-$$

$$(2.45)$$

Papain has been immobilized according to Eq. 2.46 by diazo coupling of bisdiazobenzidine with the phenolic group of tyrosine, the imidazole group of histidine or free amino groups in the enzyme protein.[96]

$$Cl^-\left[N\equiv\overset{+}{N}-\left\langle\bigcirc\right\rangle\!\!\left\langle\bigcirc\right\rangle-\overset{+}{N}\equiv N\right]Cl^- + enzyme\longrightarrow$$

$$-N=N-enzyme-N=N-\left\langle\bigcirc\right\rangle\!\!\left\langle\bigcirc\right\rangle-N=N-enzyme-N=N-$$

$$\begin{array}{c}
N \\
\| \\
N \\
| \\
\bigcirc \\
| \\
\bigcirc \\
| \\
N \\
\| \\
N \\
|
\end{array}$$

$$-N=N-enzyme-N=N-$$

$$(2.46)$$

2.1.3 Entrapping Method

The entrapping method is based on confining enzymes in the lattice of a polymer matrix or enclosing enzymes in semipermeable membranes, and can be classified into the lattice and microcapsule types. This method differs from the covalent binding and cross-linking methods in that the enzyme itself does not bind to the gel matrix or membrane. Thus, this method may have wide applicability. However, if a chemical polymerization reaction is employed for entrapping, relatively severe conditions are required and loss of enzyme activity occurs in some cases. Therefore, it is necessary to select the most suitable conditions for the immobilization of various enzymes.

A. Lattice type

The lattice-type entrapping method involves entrapping enzymes within the interstitial spaces of a cross-linked water-insoluble polymer, i.e., within the gel matrix.

Various synthetic polymers such as polyacrylamide, polyvinylalcohol and others, and natural polymers such as starch and "konjak" powder have been used for the immobilization of enzymes by this technique.

Immobilized enzymes which have been prepared by this method are listed in Table 2.10.

TABLE 2.10   Immobilization of enzymes by the lattice-type entrapping method

Material	Enzyme	Reference
Starch matrix	cholinesterase	362–364
"Konjak" powder	acid protease	365
κ-Carrageenan	aminoacylase	624
	fumarase	624
	aspartase	624
	alcohol dehydrogenase	366
	lactate dehydrogenase	192, 366, 367
	malate dehydrogenase	192
	glucose oxidase	366, 368–371
	L-amino acid oxidase	372
	D-amino acid oxidase	371
	hexokinase	370, 373
	phosphofructokinase	373
	ribonuclease	374
	cholinesterase	375, 376
	α-amylase	371, 374, 377
	β-amylase	371
	glucoamylase	371, 377–380

(Continued)

TABLE 2.10—*Continued*

Material	Enzyme	Reference
	β-glucosidase	381
	lactase	380, 382, 383
	invertase	16, 64, 371, 379, 380, 384
	trypsin	345, 370, 374, 385–389
	α-chymotrypsin	374, 385
	papain	374
	neutral protease	371
	alkaline protease	371
	asparaginase	390
	urease	304, 370
	aminoacylase	371, 391
	penicillinase	392
	apyrase	385
	citrate synthase	192, 247
	aldolase	373, 374, 393
	enolase	394
	aspartase	21
	glucose isomerase	395, 396
	glucose phosphate isomerase	373
Polyacrylic acid	glucoamylase	397
	lactase	397
	invertase	397
Polyethyleneglycol dimethacrylate	invertase	398
Polyvinyl alcohol	glucoamylase	379
	invertase	379
Polyvinyl pyrrolidone	glucoamylase	399
	lactase	399
	invertase	399, 400
Silica resin†	acetylcholinesterase	401
	trypsin	385
	α-chymotrypsin	385
	apyrase	385

†
$$\left[\begin{array}{c} CH_3 \\ | \\ -Si-O- \\ | \\ CH_3 \end{array} \right]_n$$

The cross-linked polymer most commonly employed for enzyme entrapment is polyacrylamide gel, and it has been used for the immobilization of many enzymes. The procedure for the formation of polyacrylamide gel is identical to that employed for the preparation of gel commonly used for disc electrophoresis.

$$
\begin{array}{c}
CH_2=CH \\
| \\
CO \\
| \\
NH \\
| \\
CH_2 \\
| \\
NH \\
| \\
CO \\
| \\
CH_2=CH \\
BIS
\end{array}
$$

$$
\begin{array}{c}
CH_2=CH \\
| \\
CONH_2 \\
\text{acrylamide} \\
\text{monomer}
\end{array}
\quad + \quad
\quad + \text{ enzyme} \quad \xrightarrow[\text{DMAPN}]{K_2S_2O_8}
$$

$$
\begin{array}{c}
-CH_2-CH-CH_2-CH-CH_2-CH-CH_2-CH-CH_2-CH-CH_2- \\
\quad | \qquad\quad | \qquad\quad | \qquad\quad | \qquad\qquad | \\
\quad CO \quad\quad CONH_2 \; CONH_2 \; CO \qquad\quad CONH_2 \\
\quad | \qquad\qquad\qquad\qquad\qquad | \\
\quad NH \qquad\qquad\qquad\qquad\quad NH \\
\quad | \qquad\quad (\text{enzyme}) \qquad | \\
\quad CH_2 \qquad\qquad\qquad\qquad\; CH_2 \\
\quad | \qquad\qquad\qquad\qquad\qquad | \\
\quad NH \qquad\qquad\qquad\qquad\quad NH \\
\quad | \qquad\qquad\qquad\qquad\qquad | \\
\quad CO \qquad\qquad\qquad\qquad\quad CO \\
-CH_2-CH-CH_2-CH-CH_2-CH-CH_2-CH-CH_2-CH-CH_2- \\
\quad | \qquad\qquad\quad | \qquad\qquad\qquad\quad | \\
\quad CO \qquad\quad CONH_2 \qquad\qquad CO \\
\quad | \qquad\qquad\qquad\qquad\qquad\quad | \\
\quad NH \qquad\qquad\qquad\qquad\qquad\; NH
\end{array}
\qquad (2.47)
$$

As shown in Eq. 2.47, entrapment of an enzyme is generally performed by polymerizing an aqueous solution containing acrylamide and the enzyme in the presence of N,N'-methylenebisacrylamide (BIS) as a cross-linking agent. The polymerizing reaction is initiated by potassium persulfate ($K_2S_2O_8$) or riboflavin and accelerated by β-dimethylamino-propionitrile (DMAPN), N,N,N',N'-tetramethylethylene diamine (TEMED) or alum.

This technique was first used by Bernfeld et al.[374] (1963), who entrapped trypsin, papain, amylase, ribonuclease, etc., in a gel lattice of polyacrylamide.

We have investigated in detail suitable conditions for the immobilization of enzymes such as aminoacylase, asparaginase and aspartase by using polyacrylamide gel.[21,390,391] We selected the most suitable conditions to obtain highly active, stable and mechanically strong immobilized enzymes. For example, to 3 ml of a solution containing 750 mg of acrylamide monomer and 40 mg of N,N'-methylenebisacrylamide, 1 ml of a solution of enzyme dissolved in a suitable buffer was added. To this mixture, 0.5 ml of 5% dimethylaminopropionitrile as an accelerator and 1% potassium persulfate solution as an initiator were added, and the mixture was incubated at 23°C for 10 min. Polymerization occurred and a stiff agar-like gel containing the enzyme was obtained. The pore size of the resulting polyacrylamide gel appeared to be 10~40Å. Therefore, small substrate and product molecules can transfer across the gel matrix, but large enzyme molecules cannot permeate the gel lattice.

Recently, X- or γ-radiation has been used to initiate the polymerization of acrylamide monomer.[380,387] For example, Dobo[387] carried out the immobilization of trypsin by the X-ray-induced polymerization of acrylamide monomer. Namely, 200 mg of acrylamide monomer, 120 mg of BIS and 10 mg of trypsin were dissolved in 10 ml of buffer solution. The solution in an ampoule was deaerated for 15 min by the introduction of purified argon gas. After sealing the ampoule under a vaccum, 2300 rad/h of X-rays was applied for 3 h to give the immobilized enzyme.

One of the advantages of using radiation is that the polymerization can be carried out in a frozen state, where less inactivation of the enzyme is expected to occur. Another advantage is that this method can be used to prepare various shapes of immobilized enzymes by varying the type of vessel. However, the activities of immobilized enzymes prepared so far by this method have not been significantly higher than those of the enzymes immobilized with chemical catalysts. Also, the method using radiation has some disadvantages with regard to cost and equipment, and it has not been applied for industrial purposes.

In the case of the above procedures, the polyacrylamide gel can be granulated to give suitable particles as required after the polymerization.

Recently, a procedure for preparing spherical polyacrylamide beads has been reported.[383,388,389] In this procedure an aqueous solution containing enzyme and acrylamide monomer is dispersed in a hydrophobic phase consisting of toluene and chloroform, then polymerized under N_2 gas, resulting in well-defined spherical beads.

Besides the method using polyacrylamide, entrapping of enzymes has been carried out by γ-ray irradiation of polyvinylalcohol in the cases of glucoamylase and invertase.[379] The use of a silicon resin for entrapment of acetylcholinesterase was reported.[401] A naturally occurring polymer, "konjak" powder, was used for the entrapment of acid protease.[365] A starch matrix method was also reported for entrapping as follows.[362-364] A mixture of enzyme and starch-glycerine buffer solutions was dispersed in polyurethane foam pads at 47°C and dried.

More recently, entrapment of enzymes using photocross-linkable resin has been investigated by Fukui et al.[398] For example, solid polyethyleneglycol dimethacrylate polymer of different chain lengths (53) was mixed with an initiator, benzoin ethylester, and the mixture was melted by warming at 50°C. To the melted resin, a solution of invertase dissolved in 0.1 M acetate buffer (pH 5.0) was added, and the homogeneous liquid mixture was illuminated with near-ultraviolet light for a few minutes to give immobilized enzyme.

$$\underset{O}{\overset{CH_3}{\underset{\|}{CH_2=C-C-O}}}-(CH_2CH_2-O)_{\overline{n}}\underset{O}{\overset{CH_3}{\underset{\|}{C-C}}}=CH_2$$

$$53$$

Further, entrapment of enzymes using κ-carrageenan, a kind of polysaccharide extracted from red seaweeds has been investigated by the authors.[624] For example, 100 mg of enzyme were dissolved in distilled water at 37∼50°C, and 1.7 g of κ-carrageenan were dissolved in 34 ml of physiological saline at 40∼60°C. The both solutions were mixed, and the mixture was cooled at around 10°C. In order to strengthen the gel-strength, the resulting gel was soaked to cold 0.3 M potassium chloride solution. After this treatment, the obtained stiff gel was granulated in a suitable particle size to give immobilized enzyme. Further, if the immobilized enzyme is treated with suitable hardening reagents such as tannin, glutaraldehyde or hexamethylenediamine, the more stable immobilized enzyme can be obtained. This method is applicable for immobilization of many enzymes, because the procedure is very simple and is carried out under mild conditions.

B. Microcapsule type

The microcapsule-type entrapping method involves enclosing the enzymes with semipermeable polymer membranes. The resulting enzyme microcapsules generally have a diameter of 1∼100μ.

In an application of microcapsules, National Cash Register Co. developed a carbon-less copy paper in 1954. Since then, microcapsules have been applied in many fields, for instance, medical, foods, cosmetics, dyes, fuel, etc.

For the preparation of enzyme microcapsules, extremely well-controlled conditions are required in comparison with the microencapsulation of other chemical materials.

The procedures for the microencapsulation of enzymes can be classified in three categories as follows; a) interfacial polymerization, b) liquid drying and c) phase separation.

Enzymes which have been microencapsulated are listed in Table 2.11.

TABLE 2.11 Immobilization of enzymes by the microcapsule-type entrapping method

Procedure	Material	Enzyme	Reference
Interfacial polymerization	nylon	lactase	402
		asparaginase	403–406
		urease	407–411
		carbonic anhydrase	410, 412
	polyurea	asparaginase	406

(*Continued*)

TABLE 2.11—*Continued*

Liquid drying	ethyl cellulose	lipase	413
	ladder polymer of	catalase	413
	phenylsiloxane	lipase	413
		urease	413
		zymase (yeast extract)	414
		enzyme extracted from muscle	414
	polystyrene	catalase	413
		invertase	16
		lipase	413
		urease	413
Phase separation	butyl acetate cellulose	urease	320, 415
	nitro cellulose	lactase	416
		asparaginase	417
	collodion	alcohol dehydrogenase	418
		malate dehydrogenase	418
		catalase	419–423
		pyruvate kinase	424
		hexokinase	424
		lactase	425
		asparaginase	423
		urease	411, 419–421, 423

a. Interfacial Polymerization Method

This procedure is based on enclosing enzymes in semipermeable membranes of polymers by applying the principle that hydrophobic and hydrophilic monomers polymerize at the interface.

This microencapsulation technique is shown schematically in Fig. 2.3.

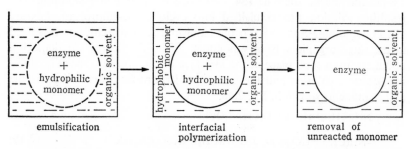

emulsification   interfacial polymerization   removal of unreacted monomer

Fig. 2.3 Preparation of microcapsules by the interfacial polymerization method.

An aqueous solution of the enzyme and a hydrophilic monomer is emulsified in a water-immiscible organic solvent. To the emulsion, hydrophobic monomer solution dissolved in the same organic solvent is added with stirring. Polymerization of both monomers occurs at the interface between the aqueous and organic solvent phases in the emulsion. By this technique, enzyme in the aqueous phase is enclosed by a membrane of polymer. Monomers reacting by condensation or addition polymerization are used for this procedure. The combinations of monomers and the resulting polymers are shown in Table 2.12.

TABLE 2.12 Combinations of monomers and the resulting polymers

Hydrophilic monomer	Hydrophobic monomer	Polymer
Polyamine	polybasic acid chloride	polyamide
	bishaloformate	polyurethane
	polyisocyanate	polyurea
Glycol, polyphenol	polybasic acid chloride	polyester
	polyisocyanate	polyurea

In this procedure, the size of capsules can be adjusted as required, and the time needed for microcapsulation is very short. However, some enzymes are unstable to the monomers used and are inactivated during the microencapsulation procedure.

Consequently, the suitable combination of monomers for microencapsulation must be selected in the light of the properties of the target enzyme and the intended application of the microencapsulated enzyme.

The microencapsulation of enzymes by this procedure was first reported by Chang et al. They microencapsulated asparaginase,[403,404] urease[407,408,410,411] and carbonic anhydrase[410,412] using a nylon membrane.

We have also microencapsulated asparaginase using nylon and polyurea membranes.[405,406] For example, asparaginase, L-asparagine and 1,6-hexamethylene diamine (54) were dissolved in buffer solution. The resulting solution was emulsified into cyclohexane-chloroform mixture (5 : 1) containing Span 85. To the emulsion, sebacoyl chloride dissolved in the same organic solvent was added (55) with stirring. Condensation polymerization took place at the interface of the organic solvent containing sebacoyl chloride and the aqueous microdroplets containing enzyme and diamine as shown in Eq. 2.48.

$$NH_2(CH_2)_6NH_2 + ClCO(CH_2)_8COCl \longrightarrow$$
$$\underset{54}{} \quad \underset{55}{}$$
$$\underset{56}{-NH(CH_2)_6NHCO(CH_2)_8CONH(CH_2)_6NHCO(CH_2)_8CO-} \qquad (2.48)$$

The resulting nylon (56) microcapsules containing asparaginase were collected by centrifugation or filtration, and sufficiently washed with ethanol and then with buffer solution. A photomicrograph of an enzyme nylon

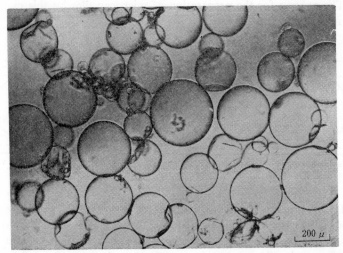

Photo 2.1 Photomicrograph of nylon microcapsules containing aspara-ginase.[405]

microcapsule is shown in Photo 2.1. When 2,4-toluenediisocyanate (**57**) was used instead of sebacoyl chloride, a polyurea membrane (**58**) was produced according to Eq. 2.49.

$$H_2N(CH_2)_6NH_2 \quad + \quad OCN-\langle \bigcirc \rangle-CH_3 \quad \longrightarrow$$
$$54 \qquad\qquad 57 \quad NCO$$

$$\left[-NH(CH_2)_6NHCONH-\langle \bigcirc \rangle-CH_3 \atop \qquad\qquad NHCONH(CH_2)_6NHCONH-\langle \bigcirc \rangle-CH_3 \atop \qquad\qquad\qquad\qquad NHCO^- \right]_n$$
$$58$$

$$(2.49)$$

In this technique, the diameter of the microcapsules can be adjusted as desired by changing the concentration of emulsifying agent (Span 85) and the speed of the mechanical emulsifier.

b. Liquid Drying Method

This method is based on enclosing enzymes via processes of dispersion of enzymes in polymer dissolved in a water-immiscible organic solvent, followed by dispersion in an aqueous solution and drying.

The process for the preparation of microcapsules by this method is shown schematically in Fig. 2.4.

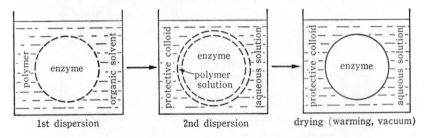

1st dispersion   2nd dispersion   drying (warming, vacuum)

Fig. 2.4   Preparation of microcapsules by the liquid drying method.

In this process, a polymer is dissolved in a water-immiscible organic solvent which has a boiling point lower than that of water. An aqueous solution of enzyme is dispersed in the organic phase to form a first emulsion of water-in-oil type. In this case, the use of oil-soluble surfactants as emulsifing agents is effective. The first emulsion containing aqueous microdroplets is then dispersed in an aqueous phase containing protective colloidal substances such as gelatin, polyvinylalcohol and surfactants, and a secondary emulsion is prepared. With continued stirring, the organic solvent is removed by warming *in vacuo*. By this procedure, a polymer membrane is produced to give enzyme microcapsules.

As polymer membranes for the microencapsulation of enzymes, organic solvent-soluble polymers such as ethyl cellulose, polystyrene, chlororubber and a ladder polymer of phenylsiloxane have been used. Commonly employed organic solvents are benzene, cyclohexane and chloroform.

The size of microcapsules can be varied by changing the concentration of polymer, the speed of the mechanical emulsifier or the kind of protective colloidal substance. However, this procedure is not suitable for the preparation of fine microcapsules with a diameter below 30 μ. Unlike in the interfacial polymerization method, preformed polymers are used and reactive reagents are not necessary in this method. Therefore, little or no inactivation of the enzyme occurs during microcapsule preparation, and this procedure seems to be very useful. However, there are several disadvantages to this method. In the process of preparing the second emulsion, the emulsion is sometimes not formed, and the yield of microcapsules becomes low. Further, it takes a long time to remove organic solvents completely, which is necessary for the formation of the solid.

The microencapsulation of catalase, lipase and zymase has been carried out by this method using ethylcellulose and polyethylene.[413]

c. Phase Separation Method

One purification method for polymers involves dissolving the polymer in an organic solvent and reprecipitating it by adding another organic solvent which is miscible with the first but which does not dissolve the polymer.

In this treatment, if the temperature and the amount of organic solvent added are varied, the polymer solution can be separated into two phases containing low and high concentrations of polymer. This phenomenon of phase separation is called "coacervation". The phase separation method for microencapsulation is an application of this phenomenon. This method is shown in Fig. 2.5.

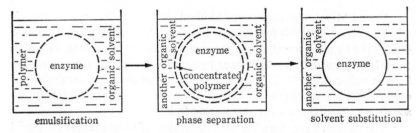

emulsification   phase separation   solvent substitution

Fig. 2.5   Preparation of microcapsules by the phase separation method.

Namely, an aqueous solution of enzyme is dispersed in a water-immiscible organic solvent containing polymer. To the organic phase containing aqueous microdroplets, another organic solvent is slowly added with stirring. The polymer is concentrated and membranes are formed around the aqueous microdroplets, enclosing the enzyme. In this procedure it is most important that polymer should not be precipitated out but separated as a phase of concentrated polymer solution. Unless this occurs, the polymer is precipitated and the enzyme is not microencapsulated.

As materials for the capsule membrane, polymers which are soluble in organic solvents are commonly used, and both organic solvents must be water-immiscible.

Table 2.13 shows examples of combinations of materials causing phase separation.

This microencapsulation is carried out under relatively mild conditions, like the liquid drying method. Therefore, this procedure is suitable for the microencapsulation of enzymes, though it has the disadvantage that the complete removal of organic solvent remaining on the polymer membrane is difficult.

Chang *et al.*[411,419-423] succeeded in the microencapsulation of aspara-

TABLE 2.13 Combinations of components causing phase separation

Polymer	Organic solvent	Second organic solvent
Ethyl cellulose	carbon tetrachloride	petroleum ether
Nitrocellulose	ether	butyl benzoate
Polystyrene	xylene	petroleum ether
Polyethylene	xylene	amyl chloride
Polyvinyl acetate	chloroform	petroleum ether
Polymethylmethacrylate	chloroform	petroleum ether
Polyisobutylene	chloroform	ethyl acetate

ginase, urease and catalase by this procedure. They also reported that a more stable microencapsulated urease was obtained when the enzyme was microencapsulated after being adsorbed on silica gel.[320,415]

d. Liquid Membrane (liposome) Method

The above methods for the microencapsulation of enzymes have been carried out by using a water-insoluble semipermeable membrane, but more recently, encapsulation of enzymes by means of a liquid membrane has been attempted.[426-430] This new procedure involves encapsulation of enzyme solution within an amphipathic liquid-surfactant membrane (liposome), e.g. formed from surfactants and lecithin.

The general procedure for immobilization of enzymes using a liquid membrane is similar to that described above.

The immobilization of glucoamylase with a liquid membrane was attempted as follows.[426] Namely, lecithin, cholesterol and dicetyl phosphate in a molar ratio of 7:2:1 were dissolved in chloroform, and an enzyme solution was added. The mixture was transferred to a rotary evaporator, and emulsified by gentle rotation under N_2 gas at 37°C. The emulsion was kept at room temperature for about 2 h and then sonicated for 10 sec under a stream of N_2 gas at 4°C. It was kept at room temperature for another 2 h and passed through a Sepharose 6B column. The liposome fractions were collected and centrifuged. The resulting pellets were suspended in buffered saline, then sonicated and centrifuged. The pellets were again suspended in buffered saline and passed through a Sepharose 6B column. Stable microcapsules consisting of a liquid membrane containing enzyme were obtained at the void volume fraction.

The most significant characteristic of liquid fraction membrane microcapsules is that the permeability of substrate and/or product across the liquid membranes is independent on the pore size of membranes but depends on the solubility of the substrate and/or product in the components of the membrane. Therefore, applications different from those of conventional microcapsules are anticipated, and research on liquid membrane microcapsules is expected to increase in importance.

2.1.4 Miscellaneous Immobilization Methods

In addition to the above methods, immobilization of enzymes has been performed through the copolymerization of enzymes with N-carboxy-amino acid anhydride.[431,432] Namely, as shown in Eq. 2.50, polymerization of reactive anhydride is initiated by condensation with free amino groups of the enzyme and yields corresponding polypeptide derivatives, that is, immobilized enzyme. By this procedure aminoacylase, α-chymotrypsin and trypsin have been immobilized.

$$ (2.50) $$

Further, lysozyme was immobilized by using N-carbonylimidazole derivatives (59) according to Eq. 2.51.[389]

$$ (2.51) $$

Namely, a copolymer (60) of styrene and p-carboxystyrene was obtained by reaction in the presence of benzoyl peroxide at 80°C. To the copolymer solution dissolved in N,N'-dimethylformamide, N,N'-carbonylimidazole was added, and the mixture was left to stand for 30 min at room temperature to give the N-carbonylimidazole derivative (59) of the copolymer. To a solution of this reactive polymer, crystalline lysozyme was added, and the mixture was then transferred into water to give particles of immobilized lysozyme.

Invertase has been immobilized by complex formation of the soluble enzyme with tannic acid, which has been used as a protein precipitating agent.[433]

More recently, in order to avoid serious allergic reaction to asparaginase used as an antitumor agent, this enzyme was entrapped in autologous red blood cells.[434]

2.1.5 Commercially Available Immobilized Enzymes and Carriers

With the development of studies on the immobilization of enzymes, various immobilized enzymes and carriers have become commercially available as tools in research and analysis. These are listed in the appended tables at the end of this book.

2.1.6 Forms of Immobilized Enzymes

The forms of immobilized enzymes can be classified into the following four types; particles, membranes (film, plate), tubes and fibers. Most immobilized enzymes are prepared in particle form for the following reasons:
1) Handling is easy.
2) Commercially available carriers suitable for enzyme immobilization are generally in particle form.
3) The surface area of particulate carriers is larger than those of other forms and the reaction efficiency is higher.

Thus, fewer papers on immobilized enzyme membranes, tubes and fibers are available in comparison with particles. Even so, these forms of immobilized enzymes are convenient for some applications. Examples of enzyme membranes, enzyme tubes and enzyme fibers are listed in Tables 2.14 and 2.15.

TABLE 2.14   Enzyme membranes

Carrier and reagent	Method	Enzyme	Reference
DEAE-cellulose sheet	carrier binding (ionic binding)	glucoamylase lactase	435 436
CM-cellulose sheet +carbodiimide	carrier binding (peptide binding)	peroxidase	437
CNBr-activated cellulose membrane	carrier binding (peptide binding)	lactase trypsin α-chymotrypsin	438 438 438
Dichloro–s-triazinyl cellulose sheet	carrier binding (alkylation)	lactase penicillin amidase	268, 274 277

(*Continued*)

TABLE 2.14—*Continued* (*1*)

Carrier and reagent	Method	Enzyme	Reference
Cellulose sheet +glutaraldehyde	carrier binding (carrier cross-linking)	peroxidase hexokinase	294 294
Collodion membrane +bisdiazobenzidine	carrier binding (carrier cross-linking)	papain	439, 440
Collodion membrane +benzidine–2,2'- disulfonic acid	carrier binding (carrier cross-linking)	papain	440, 441
Nitrocellulose membrane	entrapping (lattice)	glucose oxidase peroxidase glutamate-pyruvate transaminase aminoacylase	442 442 442 442
Cellophane membrane +glutaraldehyde	carrier binding (carrier cross-linking)	uricase lactase trypsin α-chymotrypsin	294 294 294 294
Protein-cellophane membrane+glutaralde- hyde	carrier binding (carrier cross-linking)	catalase carbonic anhydrase	443 443
Collagen membrane	carrier binding (physical adsorption)	lysozyme	444
Azide derivative of collagen membrane	carrier binding (peptide binding)	lactate dehydrogenase malate dehydrogenase glutamate dehy- drogenase aspartate amino- transferase glutamate-pyruvate transaminase hexokinase creatine kinase trypsin urease	445 445 445, 446 445, 447 445 445 445 445 445
Collagen membrane +bisdiazobenzidine	carrier binding (carrier cross-linking)	papain	448
Collagen membrane (hardening)	entrapping (lattice)	catalase lipase lactase papain urease	449 450 451 448 452–454
Collagen membrane (hardening)	electrodeposition	invertase urease	455 456
Fibrin membrane +glutaraldehyde	carrier binding (carrier cross-linking)	trypsin	457

(*Continued*)

TABLE 2.14—*Continued* (2)

Carrier and reagent	Method	Enzyme	Reference
Fibrin membrane (hardening)	entrapping (lattice)	asparaginase	458
Gelatin membrane (hardening)	entrapping (lattice)	urease	459
Protein–silicon membrane +glutaraldehyde	carrier binding (carrier cross-linking)	catalase carbonic anhydrase	443 443
Polyacrylamide gel	entrapping (lattice)	glucose oxidase catalase trypsin urease	294 294 294 294
Polyethyleneglycol dimethacrylate	entrapping (lattice)	invertase	398
Polyvinyl alcohol	entrapping (lattice)	invertase urease	442 460

TABLE 2.15 Enzyme tubes and enzyme fibers

Carrier and reagent	Method	Enzyme	Reference
ENZYME TUBES			
Diazonium derivative of nylon tube	carrier binding (diazo binding)	trypsin	461
Diazonium derivative of polyaminostyrene tube	carrier binding (diazo binding)	β-fructofuranosidase	109
Nylon tube + glutaraldehyde	carrier binding (carrier cross-linking)	alcohol dehydrogenase lactate dehydrogenase malate dehydrogenase glucose oxidase hepatic microsomal flavoprotein oxidase acetylcholinesterase glucoamylase invertase lactase trypsin urease	462 462 462 463–465 213 376 464 464 464, 466 467 463, 468
Polyacrylamide gel	entrapping (lattice)	asparaginase	469
ENZYME FIBERS			
Cellulose triacetate	entrapping (lattice)	glucose oxidase β-tyrosinase catalase lactase invertase tryptophan synthase glucose isomerase	470 470 470 470 470, 471 470 472

A. Particles

Description of the particle form is omitted in this section, as the immobilized enzymes described in the previous sections are almost all particulate.

B. Membranes (film, plate)

Enzyme membranes can be prepared by attaching enzymes to membrane-type carriers according to the methods described above, or by molding into membrane form after the enzymes have been enclosed within semipermeable membranes of polymers by the entrapping method.

For example, peroxidase was bound to ion-exchange cellulose membrane by the peptide binding method using carbodiimide[437] and lactase was bound on dichloro-*s*-triazinyl cellulose membrane by the alkylation method.[268,274] Active enzyme membranes containing uricase, lactase, trypsin and α-chymotrypsin have been prepared by the carrier cross-linking method. That is, the enzymes were impregnated on cellophane membrane, followed by treatment with glutaraldehyde as a cross-linking agent. Papain-collodion membrane was prepared by using bisdiazobenzidine as a cross-linking agent.[439,440] Lactase membrane was prepared by ionically binding the enzyme on DEAE-cellulose sheet.[436] Enzyme membrane has also been prepared by evaporating a mixture of an aqueous solution of enzyme and a solution of nitrocellulose or polyvinylalcohol dissolved in

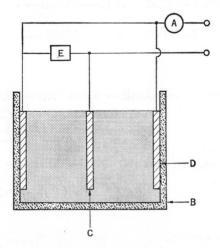

Fig. 2.6 Electrochemical apparatus for the preparation of collagen membrane.[456] A, Ammete.; B, electrolytic cell ($4 \times 7 \times 10$ cm); C, anode (platinum electrode, 6×4 cm); D, cathode (platinum electrode, 6×4 cm); E, recorder.

a mixed solvent of diethylether, ethanol, acetone, amylalcohol and ethyleneglycol.[442] Further, an electrochemical method for preparing enzyme-collagen membrane has been developed.[455,456] For example, a mixture of collagen fibrils and enzyme was placed in a cell of the type shown in Fig. 2.6. On electrolysis, enzyme-collagen membrane was formed on the surface of the platinum cathode. Enzymatically active membranes containing urease and invertase have been prepared by this technique. In this case, if a cylindrical cathode is employed and the membrane is drawn up at a constant speed, an enzyme tube can be obtained. Also, if a U-type electrode is employed, an enzyme bag can be prepared.

More recently, Fukui et al.[398] reported the preparation of enzyme membrane by using a photocross-linkable resin as described in section 2.1.3. A mixture of enzyme and photosensitive oligomers (53) was layered on a sheet of transparent polyester film, and then the layer was illuminated with near-ultraviolet light to initiate photocross-linking. Enzymes such as invertase could be entrapped in the gel lattice of the resulting membrane.

C. Tubes

Few examples of enzyme tubes are known in comparison with enzyme membranes. Nylon, polyaminostyrene and polyacrylamide tubes have been employed as carriers.

$$\sim\!\!\sim\!CONH\sim\!\!\sim\!CONH\sim\!\!\sim\!CONH\sim\!\!\sim$$

$$\xrightarrow{\text{H}^+} \sim\!\!\sim\!CONH\sim\!\!\sim\!COO^- \quad \overset{+}{H_3}N\sim\!\!\sim\!CONH\sim\!\!\sim$$

$$\xrightarrow{\text{HNO}_2} \sim\!\!\sim\!CONH\sim\!\!\sim\!COO^- \quad HO\sim\!\!\sim\!CONH\sim\!\!\sim$$

$$\xrightarrow[\text{carbodiimide}]{H_2N-\bigcirc\!\!-\!\bigcirc\!\!-NH_2} \sim\!\!\sim\!CONH\sim\!\!\sim\!C{=}O \quad HO\sim\!\!\sim\!CONH\sim\!\!\sim$$

$$\xrightarrow{\text{HNO}_2} \sim\!\!\sim\!CONH\sim\!\!\sim\!C{=}O \quad HO\sim\!\!\sim\!CONH\sim\!\!\sim$$

$$\xrightarrow{\text{enzyme}} \sim\!\!\sim\!CONH\sim\!\!\sim\!C{=}O \quad HO\sim\!\!\sim\!CONH\sim\!\!\sim \qquad (2.52)$$

For example, as shown in Eq. 2.52, the inside surface of a nylon tube was partially hydrolyzed with acid and newly liberated amino groups were destroyed by treatment with nitrous acid. The remaining carboxyl groups were converted to aminoaryl derivatives by reaction with benzidine in the presence of carbodiimide reagent. The derivative was further diazotized and enzyme was bound via diazo coupling to give an enzyme tube.[461]

The preparation of enzyme tubes by a carrier cross-linking method has also been reported. Amino groups in the enzyme and amino groups liberated by partial hydrolysis of the inside surface of a nylon tube were cross-linked with glutaraldehyde.[356,462]

D. Fibers

Many enzymes have been successfully entrapped in fibers such as cellulose triacetate.[470,472] For example, aqueous enzyme solution containing glycerol was added dropwise to cellulose triacetate dissolved in methylene chloride with gentle stirring, and emulsified. The resulting emulsion was spun into a coagulating bath containing toluene through a spinneret. The fiber was vacuum-dried to eliminate methylene chloride and toluene. By this procedure, invertase, lactase, β-tyrosinase, tryptophan synthase, glucose oxidase, catalase, etc., have been immobilized.

E. Ultrafiltration membranes

Various vessels containing a semipermeable ultrafiltration membrane (Table 2.16) have been used for continuous enzyme reactions. That is, a

TABLE 2.16   Commercially available ultrafiltration membranes

Membrane		Manufacturer	Nominal cut-off (molecular weight)
HFA	100	Abcor Inc.	10,000
	200	(U.S.A.)	20,000
	300		70,000
Diaflo	UM 05	Amicon Co.	500
	UM 2	(U.S.A.)	1,000
	UM 10		10,000
	PM 10		10,000
	PM 30		30,000
	XM 50		50,000
	XM 100A		100,000
	XM 300		300,000
PSAC	Pellicon	Millipore Ltd.	1,000
PSED	Pellicon	(U.S.A.)	25,000
PSJM	Pellicon		100,000
Diafilter	G05H	Bioengineering Co.	500
	G01T	(Japan)	1,000
	G05T		5,000
	G10T		10,000

solution of enzyme is put into an ultrafiltration vessel. The reaction is carried out within the vessel and the product is withdrawn.

Enzymes and ultrafiltration membranes which have been used for this method are listed in Table 2.17.

TABLE 2.17   Enzyme reactions using ultrafiltration membranes

Enzyme	Membrane		Reference
α-Amylase	Diaflo	PM 10	473, 474
	Diaflo	PM 30	475
β-Amylase	Diaflo	PM 30	475
Glucoamylase	Diaflo	PM 10	473, 476
	Diaflo	PM 30	475
Cellulase	HFA 300		477
Pullulanase	Diaflo	PM 30	475
α-Chymotrypsin	Diaflo	XM 100	478

As an immobilization process can be omitted, this method may be useful for some enzyme reactions. However, the application of this method is limited to cases where the substrate is high molecular and the product is low molecular, and also the enzyme itself must be stable.

In addition, continuous enzyme reactions using hollow fibers formed from the same materials as ultrafiltration membranes has been reported.[479-481] The procedure and principles of immobilization of enzymes via hollow fibers are the same as in the case of ultrafiltration membranes.

2.1.7   Conclusion

In the previous sections, methods for the preparation of immobilized enzymes have been described.

In this section, the relationship between the preparation method and the activity or characteristics of the immobilized enzyme is discussed.

A.   Method of preparation and activity of immobilized enzymes

Table 2.18 shows typical examples of the activity of immobilized enzymes prepared by various methods. In many cases, the activities of immobilized enzymes are lower than those of the native enzymes. However, in some cases of chemically modified enzymes in the soluble state, loss of activity is not seen and an increase of activity may even occur. Therefore, it may become possible in the near future to prepare immobilized enzymes having higher activities than the native enzymes.

The decrease in activity upon immobilization of an enzyme by carrier-binding methods may be ascribed to the following factors.

1)   Functional groups of essential amino acids in the active center of the enzyme participate in linking the enzyme to water-insoluble carriers.

TABLE 2.18   Preparations of immobilized enzymes and their activities

Method	Enzyme	Substrate	Relative† activity	Reference
PHYSICAL ADSORPTION				
Activated carbon	glucoamylase	starch	10	10
Acid clay	glucoamylase	starch	20	13
IONIC BINDING				
DEAE–Sephadex	aminoacylase	Ac-DL-Met	42–74	47
DEAE-cellulose	catalase	H_2O_2	70	38
	invertase	sucrose	20–70	43
	ATP deaminase	ATP	25	51
DIAZO BINDING				
Copolymer of	papain	casein	30	95
amino acids		BAEE	70	95
Aminosilanized	lactase	ONPG	75	128
porous glass	pronase	albumin	57	142
		LpNA	67	142
	L-amino acid oxidase	L-leucine	25	119
Starch-methylene dianiline compound	papain	BAEE	20–30	85
PEPTIDE BINDING				
CNBr-activated	subtilisin	casein	58–62	233
Sepharose		ATEE	79–87	233
	elastase	casein	26–27	229
		N-Ac–Ala–Ala–Ala–ME	35–37	229
CNBr-activated	glucoamylase	amylose	4	191
cellulose		maltose	60	191
Azido derivative	α-chymotrypsin	casein	1.6–2.8	157
of CM-cellulose		BTpNA	19–33	157
	trypsin	casein	1.5–3.0	157
		clupeine	8–15	157
		BAEE	15–30	157
		BApNA	94–108	157
	ficin	casein	4–5	159
		BAEE	8–12	159
	apyrase	ATP	18–33	166
Isocyanate derivative	trypsin	casein	2.5–4.0	184
of Sephadex		BAEE	17–24	184
CM-cellulose + Woodward's reagent K	apyrase	ATP	16	153
Aminosilanized porous glass + carbodiimide reagent	pepsin	hemoglobin	65	250
ALKYLATION				
Bromoacetyl	aminoacylase	Ac-DL-Met	35	264
cellulose	trypsin	casein	0.23	265

(Continued)

TABLE 2.18—*Continued*

Method	Enzyme	substrate	Relative† Activity	Reference
		LEE	0.6–0.7	265
	ribonuclease	RNA	6	265
		2′,3′-cCMP	7.5	265
Iodoacetyl cellulose	aminoacylase	Ac–DL-Met	40–50	264
Chloro–s-triazinyl cellulose	penicillin amidase	penicillin G	25	277
Copolymer of methacrylate– 4-fluorostyrene	papain	BAA	32–45	283
UGI REACTION				
Lysine–Sepharose	α-chymotrypsin	casein	20	327
		ATEE	50	327
	pepsin	hemoglobin	18–26	328
CM–Sephadex	α-chymotrypsin	casein	3	327
		ATEE	15	327
CARRIER CROSSVLINKING				
Silica + glutaralde- hyde	trypsin	casein	15	319
		BAEE	80	319
CROSS-LINKING				
Glutaraldehyde	papain	casein	12	358
		hemoglobin	16	358
		BAEE	48	358
	carboxy- peptidase A	Bz–Gly– L-Phe	30	352
Hexamethylene diisocyanate	α-chymotrypsin	ATEE	30–40	343
ENTRAPPING (LATTICE)				
Polyacrylamide gel	aminoacylase	Ac–DL-Met	50–60	391
	invertase	sucrose	8–10	384
ENTRAPPING (MICROCAPSULE)				
Nylon	asparaginase	L-Asn	37	405
Nitrocellulose	asparaginase	L-Asn	10	417
Collodione	catalase	H_2O_2	25	420

† The activity of native enzyme was taken as 100.

2) Conformational change of the enzyme occurs when the enzyme is bound to an insoluble carrier.

3) The enzyme is bound without loss of activity, but the interaction of substrate with the enzyme is affected by steric hindrance.

In the first two cases, the decrease in enzyme activity can be prevented by choosing suitable conditions for immobilization. However, the third problem is difficult to prevent unless some other immobilization method is employed.

On the other hand, in the entrapping method the decrease of enzyme activity is considered to be due to denaturation of the enzyme during the immobilization procedure, and low diffusibility of substrate and product through the gel lattice or exit membrane.

Various methods have been attempted to obtain highly active immobilized enzyme preparations. For instance, immobilization has been carried out after protecting the active center of the enzyme with a specific inhibitor, substrate or product. For example, as shown in Eq. 2.53, a sulfhydryl group of ATPase essential for its catalytic activity was first reacted with an SH-enzyme inhibitor such as *p*-hydroxymercuribenzoate, and then the protected enzyme was immobilized. The immobilized preparation was treated with a reducing agent to remove the inhibitor, and immobilized enzyme having higher activity was obtained.[482]

$$\text{enzyme-SH} \xrightarrow{\text{HOHg-}\bigcirc\text{-COOH}} \text{enzyme-SHg-}\bigcirc\text{-COOH} \xrightarrow{\text{cellulose-OCH}_2\text{CON}_3}$$

$$\underset{\text{enzyme-SHg-}\bigcirc\text{-COOH}}{\overset{\text{NHCOCH}_2\text{O-cellulose}}{|}} \xrightarrow{\text{H}^+} \underset{\text{enzyme-SH}}{\overset{\text{NHCOCH}_2\text{O-cellulose}}{|}} \qquad (2.53)$$

Moreover, immobilized lactase having higher activity was obtained by entrapping the enzyme within a lattice of polyacrylamide gel in the presence of an inhibitor such as glucono-δ-lactone or galactone-γ-lactone.[483]

We also prepared immobilized aspartase having higher activity by entrapping the enzyme within a lattice of polyacrylamide gel in the presence of its substrate, ammonium fumarate, or its product, L-aspartic acid.[21]

In addition, to obtain highly active immobilized enzyme, immobilization has been carried out as the zymogen, and then the immobilized zymogen was activated.[484] For example, immobilized chymotrypsinogen was prepared by peptide binding to succinized porous glass using carbodiimide reagent and then activated with trypsin to give immobilized chymotrypsin having higher activity.

B. Method of preparation and characteristics of immobilized enzymes

Preparation methods and characteristics of immobilized enzymes are summarized broadly in Table 2.19, though there are many exceptions.

Immobilization of enzymes by covalent binding or cross-linking is carried out under relatively severe conditions in comparison with those of

TABLE 2.19 Preparation and characteristics of immobilized enzymes

| Characteristic | Carrier binding method | | | Cross linking method | Entrapping method |
	Physical adsorption	Ionic binding	Covalent binding		
Preparation	easy	easy	difficult	difficult	difficult
Enzyme activity	low	high	high	moderate	high
Substrate specificity	unchangeable	unchangeable	changeable	changeable	unchangeable
Binding force	weak	moderate	strong	strong	strong
Regeneration	possible	possible	impossible	impossible	impossible
General applicability	low	moderate	moderate	low	high
Cost of immobilization	low	low	high	moderate	low

physical adsorption or ionic binding. Thus, in the former two cases, conformational change of the enzyme structure, and partial destruction of the active center may occur. Accordingly, unless immobilization of an enzyme by covalent binding is carried out under well-controlled conditions, immobilized enzyme having high activity cannot be obtained. However, the binding forces between the enzyme and carrier are strong, and the enzyme cannot easily be lost from carriers even in the presence of substrates or salts at high concentrations. However, when the activity of enzymes immobilized by covalent binding decreases on long-term operation, regeneration is impossible.

On the other hand, immobilization of enzymes by the ionic binding method can be achieved simply under mild conditions. Accordingly, in this case, preparations having relatively high activity are obtained. However, the binding forces between enzyme and carrier are weak in comparison with those in the covalent binding method. Therefore, leakage of the enzyme from the carrier may occur on changes of ionic strength or pH of the substrate or product solution. This type of immobilized enzyme can be regenerated when the enzyme activity decreases after prolonged operation. Thus, the ionic binding method is advantageous in comparison with the covalent binding method, particularly when expensive carriers or enzymes are used.

Unlike the ionic binding, covalent binding and cross-linking methods, in the entrapping method no binding between enzyme and carrier should occur in theory. Therefore, in many cases, preparations having high activity are obtained. However, in this method, regeneration of activity losses is impossible, as in the case of the covalent binding method. A major disadvantage of the entrapping method is that its application is limited to small molecular substrate and product molecules; the entrapped enzyme shows little or no activity toward macromolecular substrates.

Although a number of enzyme immobilization methods have been studied, no ideal general methods applicable for the immobilization of many enzymes have yet been developed. Each method has specific disadvantages. Therefore, in practice, it is necessary to find a suitable method and conditions for the immobilization of a particular enzyme in the light of the intended application.

2.2 IMMOBILIZATION METHODS FOR MICROBIAL CELLS

At present, enzymes from microorganisms are being used for industrial purposes. Microbial enzymes can be classified into two forms. One is extracellular, excreted from microbial cells during culture, and the other is intracellular, remaining in the cells. For the immobilization of the latter, the enzymes must be extracted from the cells, and in some cases purification is necessary. Extracted intracellular enzymes become unstable in many cases. If microbial cells can be immobilized directly without extraction of the enzymes, the cells can be used as a solid catalyst. In such immobilized microbial cells, the procedures for extraction and purification of enzymes can be omitted, and losses of enzyme activity may be kept to a minimum.

Further, if microbial cells having multienzyme systems are immobilized and can be used as a solid catalyst, it may be possible to replace fermentative methods involving multienzyme reactions by continuous enzyme reactions using immobilized cells. However, the following conditions will have to be satisfied; 1) the microorganism should not contain interfering enzymes catalyzing side reactions, 2) any interfering enzymes should be easily inactivated by simple methods such as heat or pH treatment, and 3) the substrates and products should pass easily through the microbial cell membrane.

On this basis, studies on the immobilization of whole microbial cells have recently been carried out. However, there are still fewer papers on immobilized microbial cells than on immobilized enzymes. From both scientific and industrial points of view, studies on immobilized microbial cells are interesting and potentially rewarding.

Three main general methods are available for the immobilization of microbial cells, as in the case of immobilized enzymes, that is, carrier-binding, cross-linking and entrapping methods. The entrapping method has been most extensively investigated so far.

2.2.1 Carrier-binding Method

This carrier-binding method is based on linking microbial cells di-

TABLE 2.20   Immobilization of microbial cells by binding to water-insoluble carriers

Water-insoluble carrier	Microorganism	Enzyme (application)	Reference
Anion exchange resin	Actinomycetes	glucose isomerase	485
Dowex 1	Azotobacter agile	multi-enzyme (oxidation of Glc, succinate)	486
	Escherichia coli		487
DEAE-cellulose	Achromobacter sp.	penicillin acylase (synthesis)	488
	Bacillus megaterium		488
Ion exchange cellulose	Achromobacter sp.	cephalosporin acylase (synthesis)	489
	Aspergillus oryzae	invertase	490
Polyvinylchloride and porous brick	yeast	multi-enzyme (production of beer)	491

rectly to water-insoluble carriers. The papers listed in Table 2.20 have been published on this method. Most of them describe ionic binding of microbial cells to water-insoluble carriers containing ion-exchange residues. This method was applied first by Hattori and Furusaka[486,487] in 1960. Living cells of *Escherichia coli* and *Azotobacter agile* bound on Dowex-1 (Cl-form) were found to have oxidative activity toward glucose and succinic acid. However, this method is considered not to be advantageous, because leakage of the enzymes may easily occur due to autolysis during the enzyme reaction.

2.2.2   Cross-linking Method

Microbial cells can be immobilized by cross-linking among the cells with bi- or multifunctional reagents, as shown in Table 2.21.

This method was used by the authors[493] in 1974 for the immobilization of *Escherichia coli* cells having high aspartase activity. We investigated the immobilization of *E. coli* cells by strengthening the cell walls or cell membranes and cross-linking the cells with a bifunctional reagent such as glutaraldehyde or toluene diisocyanate. Immobilized cells having as-

TABLE 2.21   Immobilization of microbial cells by cross-linking with bifunctional reagents

Reagent	Microorganism	Enzyme (application)	Reference
Glutaraldehyde	Bacillus coagulans	glucose isomerase	492
	Escherichia coli	aspartase (NH$_4$-fumarate→L-Asp)	493
	mold	raffinase (removal of raffinose)	

partase activity corresponding to 34.2% of that of the intact cells were obtained by using glutaraldehyde, but active immobilized cells could not be obtained with other reagents such as toluene diisocyanate.

Very few papers has so far appeared on this method. However, there is good potential if suitable cross-linking reagents for cells can be found.

2.2.3 Entrapping Method

So far, entrapping of microbial cells directly into polymer matrices

TABLE 2.22   Immobilization of microbial cells by entrapment
in polyacrylamide gel

Microorganism	Enzyme (application)	Reference
Achromobacter aceris	NAD kinase (NAD+ATP→NADP)	495
Achromobacter guttatus	hydrolysis of ε-aminocaproic acid cyclic dimer	496
Achromobacter liquidum	L-histidine ammonia-lyase (L-His→urocanate)	497
Alginomonas nonfermentas	L-menthol ester hydrolase (optical resolution)	498
Arthrobacter oxydans	multi-enzyme (1,3-propanediol→ lactic acid)	499
Brevibacterium ammoniagenes	fumarase (fumarate→L-malate)	500
	multi-enzyme (PaA, L-Cys, ATP→CoA)	501
Clostridium butyricum	multi-enzyme (Glc→H_2, hydrogen battery)	502
Corynebacterium glutamicum	multi-enzyme (Glc→L-Glu)	503
Escherichia coli	aspartase (NH_4 fumarate→L-Asp)	493
	penicillin acylase (Pen G→6-APA)	504
	tryptophan synthetase (indole+DL-Ser →L-Trp, 5-hydroxyindole+DL-Ser→ L-DOPA)	505, 506
Escherichia freundii	acid phosphatase (Glc→G–1-P, G–6-P)	507
Gluconobacter melanogenus	sorbose dehydrogenase (sorbose→sorbosone)	508
Microbacterium ammoniaphilum	diaminopimelic acid decarboxylase (DAPA→L-Lys)	509
Pseudomonas dacunhae	L-aspartate β-decarboxylase (L-Asp→L-Ala)	510
Pseudomonas putida	L-arginine deiminase (L-Arg→L-Cit)	511
Streptococcus faecalis	(Studies on L-Arg catabolism)	512
Streptomyces griseus	glucose isomerase	513
Candida tropicalis	multi-enzyme (phenol→CO_2+H_2O)	514
Curvularia lunate	steroid 11-β-hydrolase (11-deoxycortisone→cortisol)	515
Umbilica pustulata	orsellinic acid decarboxylase	516
Tetrahymena pyriformis	(confirmation of survival in polymer lattices)	517

has been most extensively investigated for the immobilization of cells. The matrices shown in Tables 2.22 and 2.23 have been employed; i) collagen, ii) gelatin, iii) agar, iv) cellulose triacetate, v) alginate, vi) polyacrylamide and vii) polystyrene. Among these materials, polyacrylamide gel has been used extensively for the immobilization of many kinds of microbial cells, as shown in Table 2.22.

A. Polyacrylamide gel

The polyacrylamide gel method was used for the immobilization of

TABLE 2.23 Immobilization of microbial cells by entrapment with other materials

Material	Microorganism	Enzyme (application)	Reference
Agar	*Lactobacillus*	β-galactosidase	518
	bulgaricus	(hydrolysis of lactose)	518
	Saccharomyces	invertase	519
	pastorianus		
Alginate	*Candida tropicalis*	multi-enzyme	
		• (phenol→CO_2+H_2O)	514
	yeast	(production of bread)	520
κ-Carrageenan	*Brevibacterium*	fumarase	624, 625
	flavum	(fumarate→L-malate)	
	Escherichia	aspartase	624, 625
	coli	(NH_4-fumarate→L-Asp)	
	Streptomyces	glucose isomerase	624, 625
	phaeochromogenes		
Cellulose	*Escherichia coli*	penicillin acylase	
triacetate		(Pen G→6-APA)	521
		(synthesis)	522
	Saccharomyces lactis	β-galactosidase	470
		(hydrolysis of lactose)	
	Streptomyces	glucose isomerase	523
	phaeochromogenes		
Collagen	*Corynebacterium*	multi-enzyme (Glc→L-Glu)	524
	glutamicum		
	soil bacteria	multi-enzyme (BOD sensor)	525
	Streptomyces	glucose isomerase	526, 527
	phaeochromogenes		
	Streptomyces venezuelae	glucose isomerase	528
Collagen and	*Erwinia herbicola*	β-tyrosinase (phenol, Pyr,	529, 530
dialdehyde starch		NH_3→Tyr; pyrocatechol,	
		Pyr, NH_3→L-DOPA)	
Egg albumin and	*Aspergillus*	aminoacylase (optical	531
glutaraldehyde	*ochraceus*	resolution of DL-Met)	
Gelatin and	Actinomycetes	glucose isomerase	532
glutaraldehyde			
Polystyrene	*Candida tropicalis*	multi-enzyme (phenol→	514
		CO_2+H_2O)	

enzymes by Bernfeld *et al.*[374] in 1963. The technique was applied to lichen cells, *Umbilicaria pustulata*, by Mosbach and Mosbach[516] in 1966.

Subsequently, we[493] extensively investigated the conditions for immobilization of bacterial cells, such as *Escherichia coli*, having high aspartase activity for industrial application. In order to prepare efficient immobilized *E. coli* cells, the type and concentration of bifunctional reagent for cross-linking and the concentration of acrylamide monomer were investigated systematically. Table 2.24 shows the results of immobilization of *E. coli* using various bifunctional reagents for lattice formation. The activities of the immobilized *E. coli* specimens obtained were almost the same except in the cases of ethylene urea and 1,3,5-triacryloyl-*s*-triazine. *N,N'*-methylenebisacrylamide (BIS) was chosen because it was commercially available at low cost. The concentrations of acrylamide monomer and BIS, and the amount of cells to be entrapped were also investigated, and optimum conditions were established for the immobilization of *E. coli*.

E. coli cells (10 kg, wet weight) collected from cultured broth were suspended in 40 liters of physiological saline. To this suspension, 7.5 kg of acrylamide monomer and 0.4 kg of BIS at 8°C were added and mixed. To the mixture, 5 liters of 5% β-dimethylaminopropionitrile, an accelerator of polymerization, and 5 liters of 2.5% potassium persulfate, an initiator of polymerization, were added, and the reaction mixture was allowed to stand at 25°C for 10∼15 min to obtain a stiff gel. In this process, the temperature of the reaction mixture tends to increase as the polymerization proceeds, but it is necessary to keep the temperature of the gel below 50°C. The resulting stiff gel was granulated to a suitable size and used for continuous enzyme reaction. Many kinds of microbial cells can be efficiently immobilized in a similar fashion.

The chemical structure of polyacrylamide gel prepared by this technique and an electron micrograph of the immobilized *E. coli* cells obtained are shown in Fig. 2.7 and Photo 2.2, respectively. The pore size of this

TABLE 2.24   Immobilization of *E. coli* cells by using various
bifunctional reagents[493]

Bifunctional reagents	Aspartase activity (μmol/h)	Yield of activity (%)
N,N'-Methylenebisacrylamide	1,220	67.0
N,N'-Propylenebisacrylamide	1,104	60.7
Diacrylamide dimethylether	1,048	57.6
1,2-Diacrylamide ethyleneglycol	1,136	62.4
N,N'-Diallyl tartardiamide	1,320	72.5
Ethylene urea bisacrylamide	128	7.0
1,3,5-Triacryloyl hexahydro–*s*-triazine	128	7.0

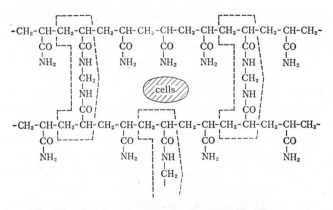

Fig. 2.7   Chemical structure of polyacrylamide gel.

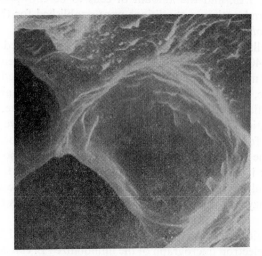

Photo 2.2   Electron micrograph of immobilized *E. coli* cells (×3000).

polyacrylamide gel varied with the concentrations of monomer and cross-linking agent, but averaged 10–40 Å under the conditions described above. Thus microbial cells, macromolecular enzymes and nucleic acids are not lost from the gel lattice during the enzyme reaction, but the low molecular substrate and product can pass freely through the gel lattice.

B.   Collagen membrane

Immobilization of microbial cells using collagen was carried out by the technique used for immobilization of enzymes developed in 1973 by Vieth[526] *et al*. This method is suitable for the preparation of membrane-type immobilized cells. In this case, the collagen membrane is weak and enzymes are readily lost from the matrix. Thus, after immobilization various hardening treatments are carried out. As hardening reagents, formaldehyde, glutaraldehyde, dialdehyde starch and so forth are employed.

C. Cellulose fibers

Immobilization of microbial cells using cellulose fibers was developed by Dinelli[470] in 1972. The enzyme activity of immobilized microbial cells prepared by this method is said to be very stable, and the technique can be applied to many enzymes or microbial cells. For example, cellulose triacetate was dissolved in methylene chloride, and a microbial cell suspension containing glycerol was added dropwise with gentle stirring. The mixture was emulsified, and the emulsion was spun through a spinneret into a coagulating bath containing toluene. The resulting fibers were dried *in vacuo* to give immobilized cell fibers. It was reported that *E. coli* cells entrapped by this method showed penicillin amidase activity amounting to about 80% of that of the intact cells, and the β-galactosidase activity of *Saccharomyces lactis* was 10% of the native value after immobilization.

D. Other entrapping methods

Entrapment of microbial cells by other methods has been reported, as shown in Table 2.23. Agar, alginate and polystyrene have been used as matrices for immobilization. However, suitable conditions for the immobilization of microbial cells and the characteristics of the resulting immobilized cells have not yet been investigated in detail.

More recently, the authors[624,625] disclosed immobilization method of microbial cells using κ-carrageenan as described in section 2.1.3. That is, microbial cell suspension and κ-carrageenan solution were mixed at 40~55°C. The mixture was cooled at around 10°C, and the resulting gel was soaked to 0.3 M potassium chloride solution in order to strengthen the gel-strength. The obtained stiff gel was granulated in a suitable size to give immobilized microbial cells. The resulting immobilized microbial cells show high enzyme activity and are stable. Further, if the immobilized microbial cells are treated with a suitable hardening reagent such as glutaraldehyde, the more stable immobilized cells can be obtained. This facile carrageenan method is applicable for immobilization of many kinds of microbial cells and is considered to be more advantageous for industrial purpose than the polyacrylamide gel method.

2.2.4 Miscellaneous Immobilization Methods

Other methods for the immobilization of microbial cells have also

TABLE 2.25 Immobilization of microbial cells by other methods

Method	Microorganism	Enzyme (application)	Reference
Aggregation by chitosan	*Streptomyces phaeochromogenes*	glucose isomerase	533
Drying after dipping in organic acid	Actinomycetes	glucose isomerase	534
Lyophilization	yeast	invertase	535
Heat treatment	Actinomycetes	glucose isomerase	536
	mold	raffinase (removal of raffinose)	494
	Pseudomonase mephitica var. *lipolytica*	lipase (hydrolysis of triacetin and tributyrin)	537
β-Irradiation	Actinomycetes	glucose isomerase	538
Liquid membrane	*Micrococcus denitrificans*	nitrate reductase	539

been reported, as shown in Table 2.25.

Fixation of enzymes within microbial cells has been carried out by heat treatment at a temperature such that the desired enzyme is not inactivated, although this method does not give positive immobilization of cells. That is, cells of *Streptomyces* sp. having glucose isomerase activity were heated at 60–85°C for 10 min.[536] By this procedure, the enzyme was fixed inside the cells and did not leak markedly from the cells when the cells were incubated under conditions suitable for the enzyme reaction. These immobilized cells were used for the industrial production of high fructose syrup from D-glucose. The glucose isomerase of *Streptomyces phaeochromogenes* was also fixed inside the cells by β-irradiation.[538]

A method for the immobilization of *Streptomyces phaeochromogenes* cells by complex formation with chitosan was also reported.[533] For example, the cells were added to chitosan solution dissolved in acetic acid. When the mixture was adjusted to pH 6.0, a complex of cells and chitosan was formed and precipitated by aggregation. The precipitated complex showed glucose isomerase activity.

Recently, immobilization of cells by means of liquid membranes has been reported.[539] In this method, a buffer solution containing cells was dispersed and emulsified in an oil phase composed of surfactants, various additives, and a hydrocarbon solvent to form encapsulated cells with liquid membranes.

2.3 IMMOBILIZED COENZYMES

Many enzymes require coenzymes to exhibit their catalytic activities. Generally, the binding force between apoenzyme and coenzyme is relatively weak, and reversible separation to give the protein moiety (apoenzyme) and coenzyme can occur easily. Recently, by utilizing this characteristic, purification of apoenzymes has been attempted using coenzymes bound to a water-insoluble carrier as described later in affinity chromatography, and several reports on the immobilization of coenzymes have been published.

As coenzymes are vitamin derivatives and their chemical structures are known, immobilization methods for them can be easily selected. However, when a coenzyme is directly bound to a water-insoluble carrier, high-molecular-weight apoenzyme often cannot approach the coenzyme closely due to the steric hindrance of the carrier. To overcome this steric hindrance, appropriate chain structures termed "spacers" or "arms" are interposed between the carrier and coenzyme. Some coenzymes which have been immobilized are described in following sections.

2.3.1 Pyridine Nucleotide Coenzymes

A number of reports have been published on the immobilization of pyridine nucleotide coenzymes, i.e., NAD and NADP (see Fig. 2.8), as shown in Table 2.26. It is clear from the table that there are many methods and carriers available for immobilization. The binding site of coenzymes to carriers is considered to be the 6-amino group of adenine, the 8-carbon atom of adenine or the hydroxyl group of ribose. Further, immobilized coenzymes obtained appear to be not homogeneous but heterogeneous preparations.

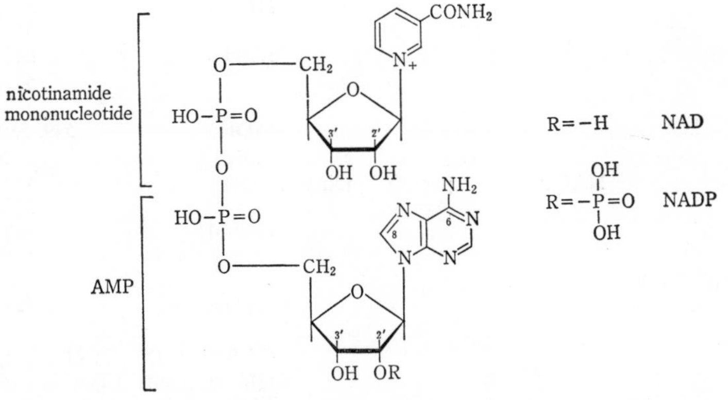

Fig. 2.8 Structures of nicotinamide adenine dinucleotide (NAD) and nicotinamide adenine dinucleotide phosphate (NADP).

TABLE 2.26   Immobilized NAD and NADP, and their applications

Carrier	Ligand	Immobilization method	Target enzyme[†]	Reference
Cellulose	NAD	CNBr activation	ADH, GDH, G3PDH, G6PDH, ICDH, LDH, MDH, TDH	540
	NADP	CNBr activation	ADH, GDH, G3PDH, G6PDH, ICDH, LDH, MDH, TDH	540
Carbonate derivative of cellulose	NAD	peptide binding	ADH, GDH, G3PDH, G6PDH, ICDH, LDH, MDH, TDH	540
Peptide derivative of cellulose	NAD	peptide binding	ADH, GDH, G3PDH, G6PDH, ICDH, LDH, MDH, TDH	540
Aromatic amino acid derivative of cellulose	NAD	diazo binding	ADH, GDH, G3PDH, G6PDH, ICDH, LDH, MDH, TDH	540
Sephadex	NAD	CNBr activation		541
Sepharose	NAD	CNBr activation	LDH	542
	N^6-(6-amino-hexyl)–NAD	CNBr activation	ADH, GDH, G3PDH, G6PDH, ICDH, LDH, MDH, TDH	543
	N^6-(N-6-aminohexyl acetamide)–NAD	CNBr activation	ADH, LDH, MDH	544–547
	8-(6-amino-hexyl)–amino-NAD	CNBr activation	LDH	548
Carbonate derivative of Sepharose	NAD	peptide binding	ADH, GDH, GTDH, ICDH, XDH	549, 550
			G3PDH	546, 547, 549–551
			G6PDH	549, 550, 552
			HBDH	549, 550, 553, 554
			LDH	542, 545–547, 549–556
			MDH	549, 550, 552
			TDH	553
	NADP	peptide binding	ADH	547, 553, 557
			G6PDH	547, 549, 552, 553, 558
			GTDH	547, 549, 553
			G3PDH, MDH, TDH	547, 553
			HBDH, ICDH, LDH, XDH	549

(*Continued*)

TABLE 2.26—*Continued*

Carrier	Ligand	Immobilization method	Target enzyme[†]	Reference
Aromatic amino acid derivative of Sepharose	NAD	diazo binding	G3PDH LDH, MDH	542, 559 542
Hydrazide derivative of Sepharose	Periodate-oxidized NAD	peptide binding		560–565
	periodate-oxidized NADP	peptide binding	G6PDH	560–565
Bromoacetyl - derivative of Sepharose	nicotinamide-6-mercapto-purine di-nucleotide	alkylation	ADH, G3PDH, LDH, MDH	566
Aromatic amino derivative of glass	NAD	diazo binding	ADH	567
Polyethyl-eneimine	succinyl NAD	peptide binding	ADH, LDH	568

† Abbreviations: AD, adenosine deaminase; ADH, alcohol dehydrogenase; cAMPK, cAMP-dependent kinase; CK, creatine kinase; CST, citrate synthase; GDH, glutamate dehydrogenase; GK, glycerate kinase; GLK, glucokinase; G3PDH, D-glyceraldehyde-3-phosphate dehydrogenase; G6PDH, D-glucose-6-phosphate dehydrogenase; GTDH, glutathion dehydrogenase; HBDH, β-hydroxylactate dehydrogenase; HK, hexokinase; ICDH, isocitrate dehydrogenase; LDH, lactate dehydrogenase; MDH, malate dehydrogenase; MIO, myosin; MK, myokinase; PFK, phosphofructokinase; 6PGDH, 6-phosphogluconate dehydrogenase; PGK, 3-phosphoglycerate kinase; PK, pyruvate kinase; PNT, pyridine nucleotide transhydrogenase; PRK, protein kinase; RNR, ribonucleotide reductase; TDH, L-threonine dehydrogenase; THD, transhydrogenase; XDH, L-xylulose dehydrogenase.

Immobilized NAD or immobilized NADP has been applied for the purification of oxidoreductases and for elucidation of the binding mechanism between coenzyme and apoenzyme, or of the reaction mechanism of the enzyme, as shown in Table 2.26. These affinity adsorbents can adsorb various enzymes, but the affinity depends on the kind of enzyme. Thus, these enzymes can be separated by using the differences of affinities. Some of these immobilized NAD(P) preparations have coenzyme activity with various oxidoreductases.

2.3.2 Adenine Nucleotides

In addition to NAD and NADP, a number of studies on the immobilization of adenine nucleotides, i.e., AMP, ADP, and ATP (see Fig. 2.9), have been reported, as listed in Table 2.27. The binding site of adenine

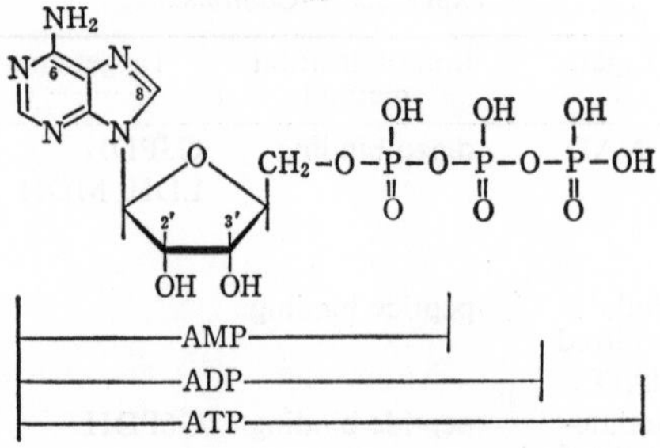

Fig. 2.9 Structures of adenine nucleotides.

TABLE 2.27 Immobilized AMP, ADP and ATP, and their applications

Carrier	Ligand	Immobilization	Target enzyme†	Reference
Sepharose	N^6-(6-amino–hexyl)-AMP	CNBr activation	ADH	543, 546, 547, 549, 550, 569–575
			CK	550, 570, 576
			G3PDH	543, 546, 547, 549–551, 569–571
			G6PDH	543, 547, 549, 550, 558, 570–573, 575, 577
			GDH	543, 547, 549, 550, 570, 571, 575
			GK	543, 550, 570–573, 576
			HBDH	543, 547, 549
			HK	543, 550, 570–573
			ICDH	543, 547, 549, 575
			LDH	543, 547, 549–551, 556, 558, 570–573, 575–579
			MDH	543, 547, 549, 550, 558, 570–573, 577
			MK	543, 550, 570, 571
			PGK	546, 550, 570, 546, 550, 570, 571

(*Continued*)

TABLE 2.27—*Continued* (*1*)

Carrier	Ligand	Immobilization	Target enzyme†	Reference
			PFK	546
			PK	546, 551
			PG6DH	575
			THD	558
	N^6-(6-amino-hexyl)-adeno-sine–2′,5′-phosphate	CNBr activation	ADH, GDH, G6PDH, ICDH, LDH	575
	N^6-(6-amino-hexyl)-ADP	CNBr activation	MIO	580, 581
	N^6-(6-amino-hexyl car-bamoyl-methyl)-ATP	CNBr activation	CST, HK	582
	ε-aminohexa-noyl-AMP	peptide binding	GK	552
	8-(6-amino-hexyl)-amino-5′-AMP	CNBr activation	G3PDH, G6PDH, MDH	577
			LDH	577, 580, 583
			PK	580
	8-(6-amino-hexyl)-amino–2′-AMP	CNBr activation	PNT	583
	8-(6-amino-hexyl)-amino–ADP	CNBr activation	MIO, GLK	580, 581
	8-(6-amino-hexyl)-amino–ATP	CNBr activation	MIO, GLK	580, 581
	8-(6-amino-hexyl)-amino–cAMP	CNBr activation	PRK, cAMPK	584, 585 / 566, 586
	thioester de-rivative of cAMP	CNBr activation	cAMPK	586
	p^1-(6-amino-hexyl)-p^2-(5′-ade-nosine)pyro-phosphate	CNBr activation	ADH, G3PDH, G6PDH, MDH	550, 571 / 550, 571, 577
			LDH	550, 571, 577, 580, 581
			PK	550, 571, 580, 581
			CK, GK, HK, MK, PGK	550, 571

(*Continued*)

TABLE 2.27—Continued (2)

Carrier	Ligand	Immobilization	Target enzyme[†]	Reference
	p-aminophe-nylester of ATP and dATP	CNBr activation	RNR	581, 587–592
Carbonate derivative of Sepharose	5′-AMP	peptide binding	ADH, GDH, G3PDH, G6PDH, HBDH, ICDH, MDH	549
			LDH	549, 556
			TDH	553
	2′-AMP	peptide binding	LDH, TDH	553, 556
	3′-AMP	peptide binding	LDH, TDH	553, 556
	ADP	peptide binding	LDH, TDH	553, 556
	ATP	peptide binding	LDH	556
	ε-aminohexa-noyl-AMP	peptide binding	LDH	556
	adenosine	peptide binding	LDH	556
	adenine	peptide binding	LDH	556
Hydrazide derivative of Sepharose	periodate-oxi-dized AMP	peptide binding		560–563, 581
	periodate-oxi-dized ADP	peptide binding		560–563, 581
	periodate-oxi-dized ATP	peptide binding	MIO	560–563, 581
	periodate-oxi-dized AMP	peptide binding	LDH, PK	577
Bromoacetyl derivative of Sepharose	6-mercapto-AMP	alkylation	AD, LDH	554
	6-mercapto–ATP	alkylation		554

[†] Abbreviations for enzymes are as in Table 2.26.

nucleotides to water-insoluble carriers is considered to be the same as in the case of pyridine nucleotides. In addition, immobilization of adenine nucleotides via phosphate has been reported. As shown in Table 2.27, these immobilized AMP, ADP and ATP preparations are useful for the purification of oxidoreductases or kinases and for clarification of the reaction mechanisms of enzymes.

2.3.3   Flavin Nucleotides

Examples of the immobilization of flavin nucleotides, i.e., FMN, FAD (see Fig. 2.10), and derivatives are listed in Table 2.28. These adsorbents are used for the purification of flavin enzymes.

Among these adsorbents, the structures of 7-cellulose acetamido-

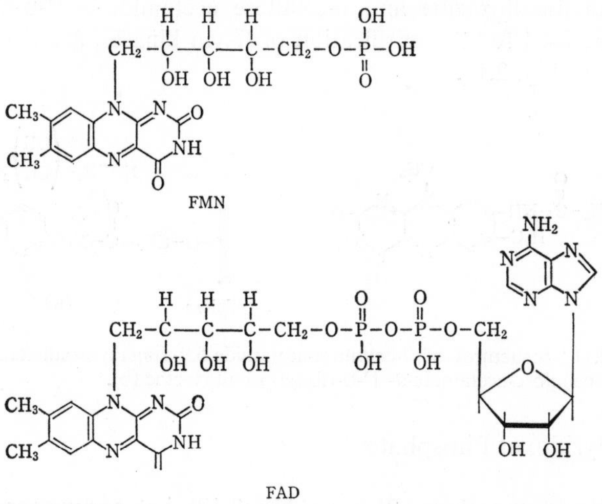

Fig. 2.10 Structures of flavin mononucleotide (FMN) and flavin adenine dinucleotide (FAD).

TABLE 2.28 Immobilized vitamin B_2 and derivatives, and their applications

Carrier	Ligand	Immobilization method	Target enzyme	Reference
Cellulose phosphate	riboflavin	POCl₃	glycolate oxidase	593
DEAE-cellulose	riboflavin	POCl₃	pyridoxine–5-phosphate oxidase	593
Cellulose	FMN	peptide binding	NADPH–cytochrome c reductase	593
Carboxy derivative of Sepharose	FMN	peptide binding		594
	FAD	peptide binding	D-amino acid oxidase	595
Amino derivative of Sepharose	3-carboxymethyl–FMN	peptide binding	apoflavoprotein	596
Chlorocarbonyl derivative of Sepharose	6-amino–9-(1′-D-ribityl)–isoalloxazine	alkylation	flavokinase	597
	7-amino–6,9-dimethyl iso-alloxazine	alkylation	flavokinase	597
Amino derivative of polyacrylamide	3-carboxymethyl-rumiflavin	peptide binding	apoflavoprotein	596

6,9-dimethylisoalloxazine and 6-cellulose acetamido-9-(1'-D-ribityl) iso-alloxazine, used for the purification of flavin kinases, are thought to be as shown in Fig. 2.11.

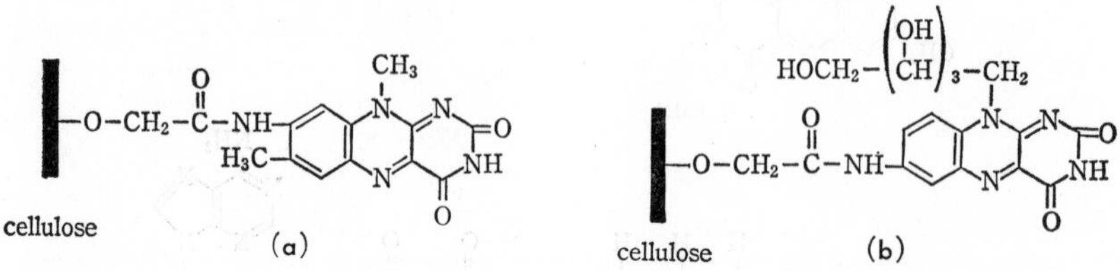

Fig. 2.11   Structures of 7-cellulose-acetamide-6,9-dimethylisoalloxazine (a), and 6-cellulose-acetamide-9-(1'-D-ribityl) isoalloxazine (b).

2.3.4   Pyridoxal Phosphate

Pyridoxal phosphate (PLP, *see* Fig. 2.12), i.e., coenzymes of vitamin B_6 enzymes, forms a Schiff base by the reaction of its 4-aldehyde group with the ε-amino group of a lysine residue of the apoenzyme. In this case,

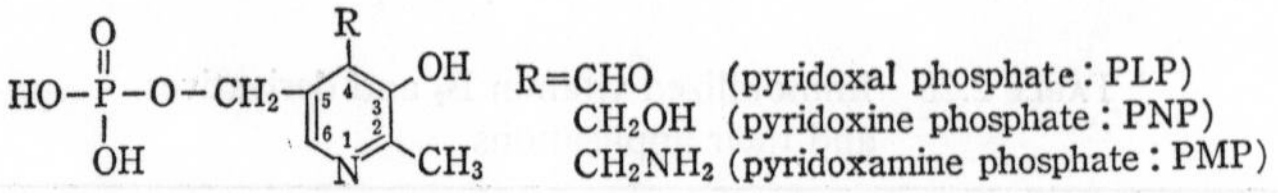

Fig. 2.12   Structures of pyridoxal phosphate, pyridoxine phosphate, and pyridoxamine phosphate.

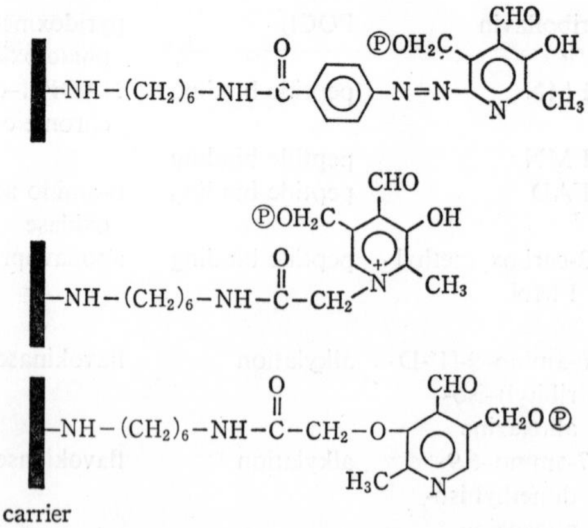

carrier

Fig. 2.13   Immobilized pyridoxal phosphates differentiated in immobilization site.[340,603] Carrier, Sepharose.
Redrawing from *op. cit.*, fig. 6-8

the 5-phosphate, 1-nitrogen atom, and 2-methyl group are thought to participate in the binding with apoenzyme. Considering these points, Fukui *et al.*[340,603] attempted the binding of PLP to Sepharose by three methods as shown in Fig. 2.13, and studied in detail the effect of the binding site of PLP on the adsorption characteristics or coenzyme activity with apoenzyme. Reports of immobilizations of PLP, PNP and PMP (see Fig. 2.12) and the applications of immobilized preparations are listed in Table 2.29.

TABLE 2.29   Immobilized PLP, PNP and PMP, and their applications

Carrier	Ligand	Immobilization method	Target enzyme	Reference
Sepharose	PMP	CNBr activation	tyrosine amino-transferase	598–600
			tyrosine amino-transferase (synthetic ribosomes)	598–600
	amino derivative of pyridoxamine	CNBr activation	glutamic-oxalo-acetic trans-aminase	601
Amino derivative of Sepharose	PLP	Schiff base (NaBH$_4$)	aspartate amino-transferase	602
Carboxy derivative of Sepharose	PMP	peptide binding	tyrosine amino-transferase	598–600
			tyrosine amino-transferase (synthetic ribosomes)	598–600
Diazonium-derivative of Sepharose	PLP	diazo binding	tryptophanase, β-tyrosinase,	340, 561 599, 603 –605
			tyrosine phenol-lyase	
			tyrosine amino-transferase (synthetic ribosomes)	
	PNP	diazo binding	tyrosine amino-transferase (synthetic ribosomes)	604
Bromoacetyl derivative of Sepharose	PLP	alkylation	tryptophanase	561, 603, 605, 606

2.3.5   Vitamin B$_{12}$

Vitamin B$_{12}$ and its derivatives (see Fig. 2.14) have also been immobilized. The immobilized preparations have been used for the purification

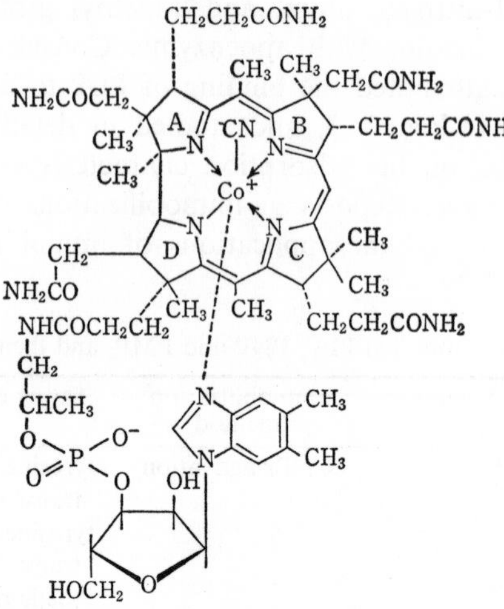

Fig. 2.14   Structure of vitamin B_{12}.

of vitamin B_{12}-related enzymes and vitamin B_{12}-binding proteins, or for elucidation of reaction mechanisms. The immobilized compounds are listed in Table 2.30.

TABLE 2.30   Immobilized vitamin B_{12} and derivatives, and their applications

Carrier	Ligand	Immobilization method	Target enzyme	Reference
Sepharose	amino derivative of B_{12} (CN)	CNBr activation	ribonucleotide reductase	607
Amino derivative of Sepharose	carboxy derivative of B_{12} (CN)	peptide binding	vitamin B_{12}-binding protein	608, 609
Bromoethyl derivative of Sepharose	B_{12} (OH)	alkylation	avidin	610, 611
Poly-L-lysine	B_{12} (OH)	peptide binding	no application	592
Poly-D-glutamate	5'-deoxyadenosyl cobalamine	peptide binding	no application	592
Succinyl-γ-globulin	5'-deoxyadenosyl cobalamine	peptide binding	no application	592

2.3.6   Other Nucleotides, Vitamins, and Derivatives

As shown in Table 2.31 and 2.32, various nucleotides, vitamins, and derivatives other than those previously described have been immobilized.

For the effective purification of coenzyme A (CoA) from microbial fermentation broth, the authors carried out the immobilization of CoA using CNBr-activated Sepharose. First, material showing specific affinity to the immobilized CoA (CoA-affinity substance) (see Fig. 2.15) was separated from microbial culture broth, and then CoA-affinity substances obtained were immobilized on CNBr-activated Sepharose. Using these

TABLE 2.31 Other immobilized nucleotides, and their applications

Carrier	Ligand	Immobilization method	Target enzyme	Reference
Sepharose	2'-deoxyuridine 5'-(6-*p*-nitro-benzamido) hexyl phosphate	diazo binding	thymidylate synthetase	612, 613
	p^1-(6-aminohexyl) $-p^2$-(5'-uridine)-pyrophosphate	CNBr activation	UDPG dehydro-genase, galactosyltrans-ferase	577, 581, 614
	5'-(4-aminohexyl) uridine-(2',3') phosphate		ribonucleases	615
Hydrazide derivative of Sepharose	periodate-oxi-dized GTP	peptide binding	dihydroneopterin phosphate syn-thetase	581, 616
			glutamate dehydrogenase	617
	periodate-oxi-dized (GMP, GTP, CMP, CTP)	peptide binding	no application	560
	periodate-oxi-dized UMP	peptide binding	no application	618

TABLE 2.32 Other immobilized vitamins and derivatives, and their applications

Carrier	Ligand	Immobilization method	Target enzyme	Reference
Sepharose	ε-*N*-biotinyllysine	CNBr activation	avidin	619
	reduced CoA	CNBr activation	CoA-affinity protein	620
Amino deriva-tive of Sepharose	thiamine pyrophosphate	peptide binding	thiamine-binding protein	621
Amino deriva-tive of polyacrylamide	N^{10}-formylamino-pterin	peptide binding	dihydrofolate reductase	622
Amino deriva-tive of glass	lipoyl chloride	alkylation	lipoamide reduc-tase	623

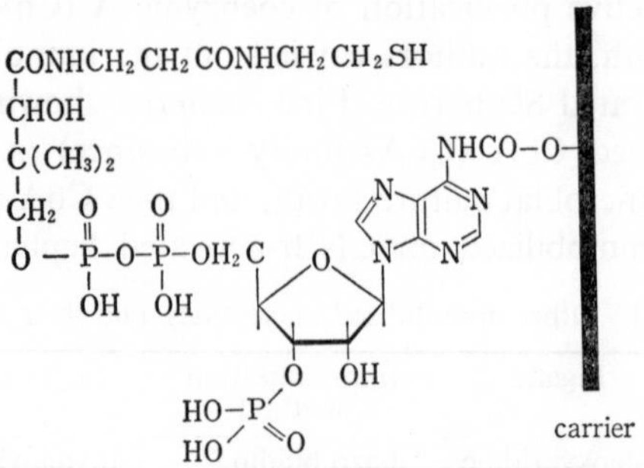

Fig. 2.15 Suggested structure for immobilized coenzyme A. Carrier, Sepharose.

immobilized CoA-affinity substances, the purification of CoA from fermentation broth could be carried out easily.

For affinity chromatography, application of this technique, i.e., isolating an affinity substance by employing immobilized pure target substance, has the advantage that unknown affinity substances in organisms can be utilized as the ligand, although the pure target substance is necessary at the initial immobilization step. This technique is effective for the separation of biological substances, especially when selection and/or preparation of the ligands are difficult.

REFERENCES

1. A. Ahmad, S. Bishayee and B. K. Bachhawat, *Biochem. Biophys. Res. Commun.*, **53**, 730 (1973)
2. J. A. Rothfus and S. J. Kennel, *Cereal. Chem.*, **47**, 140 (1970)
3. T. Watanabe, M. Fujimura, T. Mori, T. Tosa and I. Chibata, *J. Appl. Biochem.*, **1**, 28 (1979).
4. T. Watanabe, T. Mori, T. Tosa and I. Chibata, *Biotechnol. Bioeng.*, **21**, 477 (1979).
5. E. Sulkowski and M. Laskowski, Sr., *Biochem. Biophys. Res. Commun.*, **57**, 463 (1974)
6. J. Visser and M. Strating, *FEBS Letters*, **57**, 183 (1975)
7. I. Steenhoek and P. Kooiman, *Enzymologia*, **35**, 335 (1968)
8. K. Miyamoto, T. Fujii, and T. Miura, *J. Ferment. Technol.*, **49**, 565 (1971)
9. I. Stone, U.S. Patent, 2717852 (1955)
10. T. Tominaga, T. Niimi and H. Sugihara, Japanese Patent, 68-23560 (1968)
11. S. Usami, T. Yamada, and A. Kimura, *Hakko Kyokaishi* (Japanese), **25**, 513 (1967)
12. T. Tominaga, T. Niimi and H. Sugihara, Japanese Patent, 67–1360 (1967)
13. A. Kimura, H. Shirasaki and S. Usami, *Kogyokagaku Zasshi* (Japanese), **72**, 489 (1969)

14. S. Usami, M. Matsubara and J. Noda, *Hakko Kyokaishi* (Japanese), **29**, 195 (1971)
15. J. M. Nelson and E. G. Griffin, *J. Am. Chem. Soc.*, **38**, 1109 (1916)
16. S. Usami, E. Hasegawa and M. Karasawa, *Hakko Kyokaishi* (Japanese), **33**, 152 (1975)
17. B. Solomon and Y. Levin, *Biotechnol. Bioeng.*, **17**, 1323 (1975)
18. P. Monsan and G. Durand, *FEBS Letters*, **16**, 39 (1971)
19. B. J. Abbott and D. S. Fukuda, *Antimicrob. Agents and Chemther.*, **8**, 282 (1975)
20. C. Schwabe, *Biochemistry*, **8**, 795 (1969)
21. T. Tosa, T. Sato, T. Mori, Y. Matuo and I. Chibata, *Biotechnol. Bioeng.*, **15**, 69 (1973)
22. T. Toganehara and K. Watanabe, Japanese Patent, 65-27319 (1965)
23. S. Usami and N. Taketomi, *Hakko Kyokaishi* (Japanese), **23**, 267 (1965)
24. A. Traub, E. Kaufmann and Y. Teitz, *Anal. Biochem.*, **28**, 469 (1969)
25. A. D. McLaren and E. F. Estermann, *Arch. Biochem., Biophys.*, **61**, 158 (1956)
26. A. D. McLaren, *Science*, **125**, 697 (1957)
27. A. D. Traher and J. R. Kittrell, *Biotechnol. Bioeng.*, **16**, 413 (1974)
28. R. A. Messing, *Enzymologia*, **39**, 12 (1970)
29. J. P. Hummel and B. S. Anderson, *Arch. Biochem. Biophys.*, **112**, 443 (1965)
30. R. A. Messing, *Enzymologia*, **38**, 370 (1970)
31. H. L. Brockman, J. H. Law and F. J. Kézdy, *J. Biol. Chem.*, **248**, 4965 (1973)
32. R. A. Messing, *Enzymologia*, **38**, 39 (1969)
33. D. L. Marshall and J. L. Walter, *Carbohydrate Res.*, **25**, 489 (1972)
34. R. A. Messing, *Biotechnol. Bioeng.*, **16**, 897 (1974)
35. F. X. Hasselberger, B. Allen, E. K. Paruchuri, M. Charles and R. W. Coughlin, *Biochem. Biophys. Res. Commun.*, **57**, 1054 (1974)
36. T. Watanabe, Y. Matuo, T. Mori, R. Sano, T. Tosa and I. Chibata, *J. Solid-Phase Biochem.*, **3**, 161 (1978)
37. H. Maeda and H. Suzuki, *Biotechnol. Bioeng.*, **15**, 403 (1973)
38. B. P. Surinov and S. E. Manoilov, *Biochemistry* (USSR) (English Transl.), **31**, 337 (1966)
39. M. A. Mitz, *Science*, **123**, 1076 (1956)
40. M. J. Bachler, G. W. Strandberg and K. L. Smiley, *Biotechnol. Bioeng.*, **12**, 85 (1970)
41. K. L. Smiley, *ibid*, **13**, 309 (1971)
42. B. Solomon and Y. Levin, *ibid*, **16**, 1161 (1974)
43. H. Suzuki, Y. Ozawa and H. Maeda, *Agr. Biol. Chem.* (Tokyo), **30**, 807 (1966)
44. S. Usami, J. Noda and K. Goto, *Hakkokogaku Zasshi* (Japanese), **49**, 598 (1971)
45. M. A. Mitz and R. J. Schlueter, *J. Am. Chem. Soc.*, **81**, 4024 (1959)
46. A. Ya. Nikolaev, *Biochemistry* (USSR) (English Transl.), **27**, 713 (1962)
47. T. Tosa, T. Mori, N. Fuse and I. Chibata, *Enzymologia*, **31**, 214 (1966)
48. T. Tosa, T. Mori, N. Fuse and I. Chibata, *ibid*, **31**, 225 (1966)
49. T. Tosa, T. Mori, N. Fuse and I. Chibata, *ibid*, **32**, 153 (1967)
50. T. Barth and H. Masková, *Collection Czech. Chem. Commun.*, **36**, 2398 (1971)
51. S. Chung, M. Hamano, K. Aida and T. Uemura, *Agr. Biol. Chem.* (Tokyo), **32**, 1287 (1968)
52. W. Becker and E. Pfeil, *J. Am Chem. Soc.*, **88**, 4299 (1966)
53. S. Ogino, *Agr. Biol. Chem.* (Tokyo), **34**, 1268 (1970)
54. T. Tosa, T. Mori, N. Fuse and I. Chibata, *Biotechnol. Bioeng.*, **9**, 603 (1967)
55. T. Tosa, T. Mori and I. Chibata, *Agr. Biol. Chem.* (Tokyo), **33**, 1053 (1969)
56. T. Tosa, T. Mori, N. Fuse and I. Chibata, *ibid*, **33**, 1047 (1969)
57. T. Fukumura, Japanese Patent 74-15795 (1974)
58. N. Tsumura and M. Ishikawa, *Shokuhinkogyo Gakkaishi* (Japanese), **14**, 539 (1967)
59. B. H. J. Hofstee, *Biochem. Biophys. Res. Commun.*, **53**, 1137 (1973)
60. H. Brandenberger, *Rev. Ferment. Ind. Aliment.*, **11**, 237 (1956)
61. Y. K. Park, *J. Ferment. Technol.*, **52**, 140 (1974)

62. J. Boudrant and C. Ceheftel, *Biotechnol. Bioeng.*, **17**, 827 (1975)
63. A. Ya. Nikolaev and S. R. Mardashev, *Biochemistry* (USSR) (English Transl.), **26**, 565 (1961)
64. H. Nakagawa, T. Arao, T. Matsuzawa, S. Ito, N. Ogura and H. Takehana, *Agr. Biol. Chem.* (Tokyo), **39**, 1 (1975)
65. K. Miyamoto, T. Fujii, N. Tamaoki, M. Okazaki and Y. Miura, *J. Ferment. Technol.*, **51**, 566 (1973)
66. L. B. Barnett and H. B. Bull, *Biochim. Biophys. Acta*, **36**, 244 (1959)
67. W. G. Owen and R. H. Wagner, *Am. J. Physiol.* **220**, 1941 (1971)
68. W. E. Hornby, M.D. Lilly and E. M. Crook, *Biochem. J.*, **107**, 669 (1968)
69. M. A. Mitz and L. J. Summaria, *Nature*, **189**, 576 (1961)
70. M. D. Lilly, C. Money, W. E. Hornby and E. M. Crook, *Biochem. J.* **95**, 45p (1965)
71. Y. Kuriyama and F. Egami, *Seikagaku* (Japanese), **38**, 735 (1966)
72. F. Egami and Y. Kuriyama, Japanese Patent, 67-3068 (1967)
73. J. C. Lee, *Biochim. Biophys. Acta*, **235**, 435 (1971)
74. W. M. Ledingham and W. E. Hornby, *FEBS Letters*, **5**, 118 (1969)
75. R. Datta, W. Armiger and D. F. Ollis, *Biotechnol. Bioeng.*, **15**, 993 (1973)
76. K. Martinek, V. S. Goldmacher, A. M. Klibanov and I. V. Berezin, *FEBS Letters*, **51**, 152 (1975)
77. P. L.-Osheroff and R. J. Guillory, *Biochem. J.*, **127**, 419 (1972)
78. M. Tanaka, N. Nakamura and K. Mineura, Japanese Patent, 68-26286 (1968)
79. Kyowa Hakko Kogyo Kabushiki Kaisha, France Patent, 1471792 (1966)
80. I. Chibata, T. Tosa, T. Sato, T. Mori and Y. Matuo, Proc. of the IVth Int. Fermentation Symposium: Fermentation Technology Today, p. 383, Society of Fermentation Technology, Japan, 1972
81. R. D. Falb, J. Lynn and J. Shapira, *Experientia*, **28**, 958 (1973)
82. S. Kinoshita, M. Tanaka and N. Nakamura, Japanese Patent, 66-13785 (1966)
83. S. A. Barker and P. J. Somers, *Carbohydrate Res.*, **14**, 287 (1970)
84. N. E. Franks, *Biotechnol. Bioeng.*, **14**, 1027 (1972)
85. L. Goldstein, M. Pecht, S. Blumberg, D. Atlas and Y. Levin, *Biochemistry*, **9**, 2322 (1970)
86. C. B. Anfinsen, M. Sela and J. P. Cooke, *J. Biol. Chem.*, **237**, 1825 (1962)
87. I. H. Silman, D. Wellner and E. Katchalski, *Israel J. Chem.*, **1**, 65 (1963)
88. E. Katchalski, *Polyamino Acids, Polypeptides and Proteins* (ed. M. A. Stahmann), p. 283, The Univ. of Wisconsin Press, 1962
89. E. Katchalski and A. Bar-Eli, Israel Patent 13950 (1960)
90. E. Katchalski and A. Bar-Eli, *Nature*, **188**, 856 (1960)
91. A. N. Glazer, A. Bar-Eli and E. Katchalski, *J. Biol. Chem.*, **237**, 1832 (1962)
92. A. Bar-Eli and E. Katchalski, *ibid.*, **238**, 1690 (1963)
93. B. Alexander, A. Rimon and E. Katchalski, *Federation Proc.*, **24**, 804 (1965)
94. A. Rimon, B. Alexander and E. Katchalski, *Biochemistry*, **5**, 792 (1966)
95. J. J. Cebra, D. Givol, H. I. Silman and E. Katchalski, *J. Biol. Chem.*, **236**, 1720 (1961)
96. I. H. Silman, M. Albu-Weissenberg and E. Katchalski, *Biopolymers*, **4**, 441 (1966)
97. J. J. Cebra, *J. Immunology*, **92**, 977 (1964)
98. A. M. Engel and B. Alexander, *J. Biol. Chem.*, **246**, 1213 (1971)
99. A. Rimon, M. Gutman and S. Rimon, *Biochim. Biophys. Acta*, **73**, 301 (1963)
100. M. Gutman and A. Rimon, *Can. J. Biochem.*, **42**, 1339 (1964)
101. P. Cresswell and A. R. Sanderson, *Biochem. J.*, **119**, 447 (1970)
102. E. Riesel and E. Katchalski, *J. Biol. Chem.*, **239**, 1521 (1964)
103. S. A. Barker and P. J. Somers, *Carbohydrate Res.*, **14**, 323 (1970)
104. H. Mašková, T. Barth, B. Jírovský and I. Rychlik, *Collection Czech. Chem. Commun.* **38**, 943 (1973)
105. H. Brandenberger, *Angew. Chem.*, **67**, 661 (1955)
106. N. Grubhofer and L. Schleith, *Naturwissenshaften*, **40**, 508 (1953)

107. N. Grubhofer and L. Schleith, *Hoppe-Seyler's Z. Phys. Chem.*, **297**, 108 (1954)
108. T. Fukushi and T. Isemura, *J. Biochem.*, **64**, 283 (1968)
109. H. Filippusson and W. E. Hornby, *Biochem. J.*, **120**, 215 (1970)
110. S. Kudo and H. Kushiro, Japanese Patent, 68-15401 (1968)
111. H. D. Brown, A. B. Patel, S. K. Chattopadhyay and S. N. Pennington, *Enzymologia*, **35**, 233 (1968)
112. G. Manecke and G. Gunzel, *Naturwissenschaften*, **54**, 531 (1967)
113. L. Goldstein, *Biochim. Biophys. Acta*, **315**, 1 (1973)
114. J. E. Dixon, F. E. Stolzenbach, J. A. Berenson and N. O. Kaplan, *Biochem. Biophys. Res. Commun.*, **52**, 905 (1973)
115. T. L. Newirth, M. A. Diegelman, E. K. Pye and R. G. Kallen, *Biotechnol. Bioeng.*, **15**, 1089 (1973)
116. C. T. Lee and N. O. Kaplan, *Israel J. Chem.*, **12**, 529 (1974)
117. M. K. Weibel and H. J. Bright, *Biochem. J.*, **124**, 801 (1971)
118. M. K. Weibel, W. Dritschilo, H. J. Bright and A. E. Humphrey, *Anal. Biochem.*, **52**, 402 (1973)
119. H. H. Weetall and G. Baum, *Biotechnol. Bioeng.*, **12**, 399 (1970)
120. H. H. Weetall and G. Baum, Abstr. Papers, *Am. Chem. Soc.*, No. 158, Biol., 153 (1969)
121. J. C. Venter, B. R. Venter, J. E. Dixon and N. O. Kaplan, *Biochem. Med.*, **12**, 79 (1975)
122. D. A. Lappi, F. E. Stolzenbach and N. O. Kaplan, *Biochem. Biophys. Res. Commun.*, **69**, 878 (1976)
123. M. J. Grove, G. W. Strandberg and K. L. Smiley, *Biotechnol. Bioeng.*, **13**, 709 (1971)
124. G. Baum, F. B. Ward and H. H. Weetall, *Biochim. Biophys. Acta*, **268**, 411 (1972)
125. H. H. Weetall, *Nature*, **223**, 959 (1969)
126. A. R. Neurath and H. H. Weetall, *FEBS Letters*, **8**, 253 (1970)
127. D. R. Marsh, Y. Y. Lee and G. T. Tsao, *Biotechnol. Bioeng.*, **15**, 483 (1973)
128. J. H. Woychik and M. V. Wondolowski, *Biochim. Biophys. Acta*, **289**, 347 (1972)
129. L. E. Wierzbicki, V. H. Edwards and F. V. Kosikowsky, *J. Food Sci.*, **38**, 1070 (1973)
130. E. S. Okos and W. J. Harper, *ibid.*, **39**, 88 (1974)
131. L. E. Wierzbicki, V. H. Edwards and F. V. Kosikowsky, *Biotechnol. Bioeng.*, **16**, 397 (1974)
132. M. V. Wondolowski and J. H. Woychik, *ibid.*, **16**, 1633 (1974)
133. R. D. Mason and H. H. Weetall, *ibid.*, **14**, 637 (1972)
134. G. P. Royer and J. P. Andrews, *J. Biol. Chem.*, **248**, 1807 (1973)
135. H. H. Weetall, *Biochim. Biophys. Acta*, **212**, 1 (1970)
136. H. H. Weetall, *Science*, **166**, 615 (1969)
137. M. Valaris and W. J. Harper, *J. Food Sci.*, **38**, 481 (1973)
138. E. C. Lee, G. F. Senyk and W. F. Shipe, *J. Food Sci.*, **39**, 927 (1974)
139. E. C. Lee, G. F. Senyk and W. F. Shipe, *ibid.*, **39**, 1124 (1974)
140. K. Tanizawa and M. L. Bender, *J. Biol. Chem.*, **249**, 2130 (1974)
141. H. H. Weetall and R. D. Mason, *Biotechnol. Bioeng.*, **15**, 455 (1973)
142. G. P. Royer and G. M. Green, *Biochem. Biophys. Res. Commun.* **44**, 426 (1971)
143. R. D. Mason, C. C. Detar and H. H. Weetall, *Biotechnol. Bioeng.*, **17**, 1019 (1975)
144. H. H. Weetall and L. S. Hersh, *Biochim. Biophys. Acta*, **185**, 464 (1969)
145. J. Shapira, C. L. Hanson, J. M. Lyding and P. J. Reilly, *Biotechnol. Bioeng.*, **16**, 1507 (1974)
146. G. W. Strandberg and K. L. Smiley, *ibid.*, **14**, 509 (1972)
147. H. H. Weetall, *J. Biomed. Mater. Res.*, **4**, 597 (1970)
148. D. H. Campbell, E. Luescher and L. S. Lerman, *Proc. Natl. Acad. Sci.*, **37**, 575 (1951)
149. H. Leuchs, *Chem. Ber.*, **39**, 857 (1906)

150. W. Brümmer, N. Hennrich, M. Klockow, H. Lang and H. D. Orth, *Eur. J. Biochem.*, **25**, 129 (1972)
151. C. J. Epstein and C. B. Anfinsen, *J. Biol. Chem.*, **237**, 2175 (1962)
152. F. Egami and Y. Kuriyama, Japanese Patent, 67-3069 (1967)
153. A. B. Patel, S. N. Pennington and H. D. Brown, *Biochim. Biophys. Acta*, **178**, 626 (1969)
154. J. Christison, *Chem. Ind.*, **4**, 215 (1972)
155. H. Maeda and H. Suzuki, *Nogeikagaku Kaishi* (Japanese), **44**, 547 (1970)
156. N. W. H. Cheethan and G. N. Richards, *Carbohydrate Res.*, **30**, 99 (1973)
157. T. Takami and E. Ando, *Seikagaku* (Japanese) **40**, 749 (1968)
158. V. I. Surovtsev, L. V. Kozlov and V. K. Antonov, *Biochemistry* (USSR) (English Transl.), **36**, 167 (1971)
159. W. E. Hornby, M. D. Lilly and E. M. Crook, *Biochem. J.*, **98**, 420 (1966)
160. M. D. Lilly, W. E. Hornby and E. M. Crook, *ibid.*, **100**, 718 (1966)
161. C. W. Wharton, E. M. Crook and K. Brocklehurst, *Eur. J. Biochem.*, **6**, 565 (1968)
162. C. W. Wharton, E. M. Crook and K. Brocklehurst, *ibid.*, **6**, 572 (1968)
163. C. M. Ambrus, J. L. Ambrus, O. A. Roholt, B. K. Meyer and R. R. Shields, *J. Med.*, **3**, 270 (1972)
164. A. B. Patel, R. O. Stasiw, H. D. Brown and C. A. Ghiron, *Biotechnol. Bioeng.*, **14**, 1031 (1972)
165. H. D. Brown, S. K. Chattopadhyay and A. B. Patel, *Enzymologia*, **32**, 205 (1967)
166. K. P. Wheeler, B. A. Edwards and R. Whittam, *Biochim. Biophys. Acta*, **191**, 187 (1969)
167. H. D. Brown, A. B. Patel, S. K. Chattopadhyay and S. N. Pennington, *Enzymologia*, **35**, 215 (1968)
168. Y. Ohno and M. Stahmann, *Macromolecules*, **4**, 350 (1971)
169. M. K. Weibel, R. Barrios, R. Delotto and A. E. Humphrey, *Biotechnol. Bioeng.*, **17**, 85 (1975)
170. H. R. Schreiner, U.S. Patent 3282702 (1966)
171. R. A. Zingaro and M. Uziel, *Biochim. Biophys. Acta*, **213**, 371 (1970)
172. L. Goldstein, A. Lifshitz and M. Sokolovsky, *Int. J. Biochem.*, **2**, 448 (1971)
173. Y. Levin, M. Pecht, L. Goldstein and E. Katchalski, *Biochemistry*, **3**, 1905 (1964)
174. E. B. Ong, Y. Tsang and G. E. Perlmann, *J. Biol. Chem.*, **241**, 5661 (1966)
175. H. Tschesche, *Hoppe-Seyler's Z. Physiol. Chem.*, **348**, 1216 (1967)
176. K. Hochstrasser, M. Muss and E. Werle, *ibid.*, **348**, 1337 (1967)
177. H. Fritz, M. Hutzel and E. Werle, *ibid.*, **348**, 950 (1967)
178. H. Fritz, H. Schult, M. Hutzel and M. Wiedemann, *ibid.*, **348**, 308 (1967)
179. L. Goldstein, Y. Levin and E. Katchalski, *Biochemistry*, **3**, 1913 (1964)
180. E. Katchalski, *Ind. Eng. Chem.*, **57**, 14 (1965)
181. L. Goldstein, *Anal. Biochem.*, **50**, 40 (1972)
182. S. Lowey, L. Goldstein, C. Cohen and S. M. Luck, *J. Mol. Biol.*, **23**, 287 (1967)
183. A. Conte and K. Lehmann, *Hoppe-Seyler's Z. Physiol. Chem.*, **352**, 533 (1971)
184. R. Axén and J. Porath, *Nature*, **210**, 367 (1966)
185. H. Brandenberger, *Helv. Chim. Acta*, **40**, 61 (1957)
186. M. P. Coughlan and D. B. Johnson, *Biochim. Biophys. Acta*, **302**, 200 (1973)
187. V. R. Srinivasan and M. W. Bumm, *Biotechnol. Bioeng.*, **16**, 1413 (1974)
188. P. J. Robinson, P. Dunnill and M. D. Lilly, *ibid.*, **15**, 603 (1973)
189. J. C. Smith, I. J. Stratford, D. W. Hutchinson and H. J. Brentnall, *FEBS Letters*, **30**, 246 (1973)
190. C. H. Hoffman, E. Harris, S. Chodroff, S. Michelson, J. W. Rothrock, E. Peterson and W. Reuter, *Biochem. Biophys. Res. Commun.*, **41**, 710 (1970)
191. H. Maeda and H. Suzuki, *Agr. Biol. Chem.* (Tokyo), **36**, 1839 (1972)
192. P. A. Srere, B. Mattiasson and K. Mosbach, *Proc. Nat. Acad. Sci.*, **70**, 2534 (1973)
193. B. Mattiasson and K. Mosbach, *Biochim. Biophys. Acta*, **235**, 253 (1971)
194. R. Bohnensack, W. Augustin and E. Hofmann, *Experientia*, **25**, 348 (1969)

195. T. Horigome, H. Kasai and T. Okuyama, *J. Biochem.*, **75**, 299 (1974)
196. J. Lasch, M. Iwig and H. Hanson, *Eur. J. Biochem.*, **27**, 431 (1972)
197. A. W. Burgess, L. L. Weinstein, D. Gabel and H. A. Scheraga, *Biochemistry*, **14**, 197 (1975)
198. D. Gabel, *Eur. J. Biochem.*, **33**, 348 (1973)
199. J. Carlsson, D. Gabel and R. Axén, *Hoppe-Seyler's Z. Physiol. Chem.*, **353**, 1850 (1972)
200. R. Axén, J. Porath and S. Ernback, *Nature*, **214**, 1302 (1967)
201. R. Axén, P.-Å, Myrin and J.-C. Janson, *Biopolymers*, **9**, 401 (1970)
202. S. Ikeda and S. Fukui, *FEBS Letters*, **41**, 216 (1974)
203. A. S. Levi, *ibid.*, **52**, 278 (1975)
204. S. Ikeda, Y. Sumi and S. Fukui, *ibid.*, **47**, 295 (1974)
205. A. E. Chung, *Arch. Biochem. Biophys.*, **152**, 125 (1972)
206. S. Gestrelius, B. Mattiasson and K. Mosbach, *Biochim. Biophys. Acta*, **276**, 339 (1972)
207. L. Havekes, F. Bückmann and J. Visser, *ibid.*, **334**, 272 (1974)
208. T. Tosa, R. Sano and I. Chibata, *Agr. Biol. Chem.* (Tokyo), **38**, 1529 (1974)
209. A. E. Chung and L. E. Middleditch, *Arch. Biochem. Biophys.*, **152**, 539 (1972)
210. S. Henry, J. Koczan and T. Richardson, *Biotechnol. Bioeng.*, **16**, 289 (1974)
211. S. Grossman, M. Trop, P. Budowski, M. Perl and A. Pinsky, *Biochem. J.*, **127**, 909 (1972)
212. O. R. Zaborsky and J. Ogletree, *Biochim. Biophys. Acta*, **289**, 68 (1972)
213. S. S. Sofer, D. M. Ziegler and R. P. Popovich, *Biochem. Biophys. Res. Commun.*, **57**, 183 (1974)
214. W. W.-C. Chan, *Eur. J. Biochem.*, **40**, 533 (1973)
215. K. Feldmann, H. Zeisel and E. Helmreich, *Proc. Nat. Acad. Sci.*, **69**, 2278 (1972)
216. K. Mosbach and S. Gestrelius, *FEBS Letters*, **42**, 200 (1974)
217. I. Parikh, D. W. MacGlashan and C. Fenselau, *J. Med. Chem.*, **19**, 296 (1976)
218. S. Litvak, L. T.-Litvak, D. S. Carre and F. Chapeville, *Eur. J. Biochem.*, **24**, 249 (1971)
219. S. Murao, M. Inui and M. Arai, *Hakkokogaku* (Japanese), **55**, 75 (1977)
220. F. E. A. van Houdenhoven, P. J. G. M. de Wit and J. Visser, *Carbohydrate Res.*, **34**, 233 (1974)
221. V. Shepherd and R. Montgomery, *Biochim. Biophys. Acta*, **429**, 884 (1976)
222. R. Koelsch, *Enzymologia*, **42**, 257 (1972)
223. H. P. J. Bennett, D. F. Elliott, B. E. Evans, P. J. Lowry and C. McMartin, *Biochem. J.*, **129**, 695 (1972)
224. B. Hofsten, G. Nässén-Puu and I. Drevin, *FEBS Letters*, **40**, 302 (1974)
225. T. Seki, T. A. Jenssen, Y. Levin and E. G. Erdös, *Nature*, **225**, 864 (1970)
226. D. Gabel, P. Vretblad, R. Axén and J. Porath, *Biochim. Biophys. Acta*, **214**, 561 (1970)
227. M. L. Green and G. Crutchfield, *Biochem. J.*, **115**, 183 (1969)
228. V. Kasche, H. Lundqvist, R. Bergman and R. Axén, *Biochem. Biophys. Res. Commun.*, **45**, 615 (1971)
229. L. Sundberg and T. Kristiansen, *FEBS Letters*, **22**, 175 (1972)
230. K. S. Stenn and E. R. Blout, *Biochemistry*, **11**, 4502 (1972)
231. D. G. Deutsch and E. T. Mertz, *J. Med.*, **3**, 224 (1972)
232. B. Wiman and P. Wallen, *Eur. J. Biochem.*, **36**, 25 (1973)
233. B. Svensson, *Compt. Rend. Trav. Lab. Carlsberg*, **39**, 1 (1972)
234. H. Sekine, *Agr. Biol. Chem.* (Tokyo), **37**, 437 (1973)
235. W. W.-C. Chan, *Biochem. Biophys. Res. Commun.*, **41**, 1198 (1970)
236. W. W.-C. Chan and H. M. Mawer, *Arch. Biochem. Biophys.*, **149**, 136 (1972)
237. J. S. Mort, D. K. K. Chong and W. W.-C. Chan, *Anal. Biochem.*, **52**, 162 (1973)
238. E. A. Wider de Xifra, S. Mendiara and A. M. delC Batlle, *FEBS Letters*, **27**, 275 (1972)

239. A.-C. Johansson and K. Mosbach, *Biochim. Biophys. Acta*, **370**, 339 (1974)
240. J. F. Kennedy and A. Zamir, *Carbohydrate Res.*, **29**, 497 (1973)
241. C. J. Gray and T. H. Yeo, *ibid.*, **27**, 235 (1973)
242. T. Wagner, C. J. Hsu and G. Kelleher, *Biochem. J.*, **108**, 892 (1968)
243. J. L. Garnett, R. S. Kenyon and M. J. Liddy, *J. C. S. Chem. Commun.*, 735 (1974)
244. N. Weliky, F. S. Brown and E. C. Dale, *Arch. Biochem. Biophys.*, **131**, 1 (1969)
245. S. Shaltiel, R. Mizrahi, Y. Stupp and M. Sela, *Eur. J. Biochem.*, **14**, 509 (1970)
246. K. Mårtensson, *Biotechnol. Bioeng.*, **16**, 579 (1974)
247. K. Mosbach and B. Mattiasson, *Acta Chem. Scand.*, **24**, 2084 (1970)
248. J. B. Taylor and H. E. Swaisgood, *Biochim. Biophys. Acta*, **284**, 268 (1972)
249. S. W. Sae, *Biotechnol. Bioeng.*, **16**, 275 (1974)
250. W. F. Line, A. Kwong and H. H. Weetall, *Biochim. Biophys. Acta*, **242**, 194 (1971)
251. M. Valaris and W. J. Harper, *J. Food Sci.*, **38**, 477 (1973)
252. H. R. Horton, H. E. Swaisgood and K. Mosbach, *Biochim. Biophys. Res. Commun.*, **61**, 1118 (1974)
253. P. V. Sundaram, *ibid.*, **61**, 667 (1974)
254. H. D. Brown, C. W. Barker and J. K. Vincent, *Biotechnol. Bioeng.*, **16**, 1425 (1974)
255. G. J. Bartling, S. K. Chattopadhyay, C. W. Barker and H. D. Brown, *Can. J. Biochem.*, **53**, 868 (1975)
256. R. Axén and S. Ernback, *Eur. J. Biochem.*, **18**, 351 (1971)
257. B. Svensson, *FEBS Letters*, **29**, 167 (1973)
258. R. Jost, T. Miron and M. Wilchek, *Biochim. Biophys. Acta*, **362**, 75 (1974)
259. M. Wilchek, T. Oka and Y. J. Topper, *Proc. Nat. Acad. Sci.*, **72**, 1055 (1975)
260. S. A. Barker, H. Cho Tun, S. H. Doss, C. J. Gray and J. F. Kennedy, *Carbohydrate Res.*, **17**, 471 (1971)
261. G. Manecke, G. Günzel and H. J. Förster, *J. Polymer Sci.*, part C, No. 30, 607 (1970)
262. W. M. Herring, R. L. Laurence and J. R. Kittrell, *Biotechnol. Bioeng.*, **14**, 975 (1972)
263. H. H. Weetall and L. S. Hersh, *Biochim. Biophys. Acta*, **206**, 54 (1970)
264. T. Sato, T. Mori, T. Tosa and I. Chibata, *Arch. Biochem. Biophys.*, **147**, 788 (1971)
265. A. Patchornik, Israel Patent, 18207 (1965)
266. H. Maeda and H. Suzuki, *Agr. Biol. Chem.* (Tokyo), **36**, 1581 (1972)
267. J. R. Wykes, P. Dunnill and M. D. Lilly, *Biotechnol. Bioeng.*, **17**, 51 (1975)
268. G. Kay, M. D. Lilly, A. K. Sharp and R. J. H. Wilson, *Nature*, **217**, 641 (1968)
269. R. J. H. Wilson, G. Kay and M. D. Lilly, *Biochem. J.*, **108**, 845 (1968)
270. J. R. Wykes, P. Dunnill and M. D. Lilly, *Nature*, **230**, 187 (1971)
271. D. G. Knorre, N. V. Melamed, V. K. Starostina and T. N. Shubina, *Biochemistry* (USSR) (English Transl.), **38**, 101 (1973)
272. R. J. H. Wilson and M. D. Lilly, *Biotechnol. Bioeng.*, **11**, 349 (1969)
273. S. P. O'neill, P. Dunnill and M. D. Lilly, *ibid.*, **13**, 337 (1971)
274. A. K. Sharp, G. Kay and M. D. Lilly, *ibid.*, **11**, 363 (1969)
275. G. Kay and E. M. Crook, *Nature*, **216**, 514 (1967)
276. G. Kay and M. D. Lilly, *Biochim. Biophys. Acta*, **198**, 276 (1970)
277. D. A. Self, G. Kay, M. D. Lilly and P. Dunnill, *Biotechnol. Bioeng.*, **11**, 337 (1969)
278. D. Warburton, K. Balasingham, P. Dunnill and M. D. Lilly, *Biochim. Biophys. Acta*, **284**, 278 (1972)
279. W. H. Stimson and A. Serafini-Fracassini, *FEBS Letters*, **17**, 318 (1971)
280. G. Manecke and G. Günzel, *Makromol. Chem.*, **51**, 199 (1962)
281. G. Manecke, *Pure Appl. Chem.*, **4**, 507 (1962)
282. G. Manecke and S. Singer, *Makromol. Chem.*, **39**, 13 (1960)
283. G. Manecke and H. J. Förster, *ibid.*, **91**, 136 (1966)
284. E. Brown and A. Racois, *Bull. Soc. Chim., France*, **12**, 4351 (1971)
285. E. Brown and A. Racois, *ibid.*, **12**, 4357 (1971)

286. E. Brown, A. Racois and H. Gueniffey, *Tetrahedron Letters*, 25, 2139 (1970)
287. D. L. Regan, P. Dunnill and M. D. Lilly, *Biotechnol. Bioeng.*, 16, 333 (1974)
288. A. P. Ryle, *Int. J. Peptide Protein Res.*, 4, 123 (1972)
289. C. K. Glassmeyer and J. D. Ogle, *Biochemistry*, 10, 786 (1971)
290. A. F. S. A. Habeeb, *Arch. Biochem. Biophys.*, 119, 264 (1967)
291. J. Balcom, P. Foulkes, N. F. Olson and T. Richardson, *Process Biochem.*, 6, 42 (1971)
292. W. L. Stanley, G. G. Watters, B. G. Chan and J. M. Mercer, *Biotechnol. Bioeng.*, 17, 315 (1975)
293. W. L. Stanley, G. G. Watters, S. H. Kelly, B. G. Chan, J. A. Garibaldi and J. E. Schade, *ibid.*, 18, 439 (1976)
294. G. Broun, D. Thomas, G. Gellf, D. Domurado, A. M. Berjonneau and C. Guillon, *ibid.*, 15, 359 (1973)
295. Y. Takasaki, *Agr. Biol. Chem.* (Tokyo), 38, 1081 (1974)
296. P. D. Weston and S. Avrameas, *Biochem. Biophys. Res. Commun.*, 45, 1574 (1971)
297. F. Widmer, J. E. Dixon and N. O. Kaplan, *Anal. Biochem.*, 55, 282 (1973)
298. N. N. Rehak, J. Everse, N. O. Kaplan and R. L. Berger, *ibid.*, 70, 381 (1976)
299. D. L. Marshall, *Biotechnol. Bioeng.*, 15, 447 (1973)
300. H. H. Weetall and C. C. Detar, *ibid.*, 16, 1095 (1974)
301. D. R. Marsh and G. T. Tsao, *ibid.*, 18, 349 (1976)
302. H. H. Weetall, N. B. Havewala, W. H. Pitcher, Jr., C. C. Detar, W. P. Vann and S. Yaverbaum, *ibid.*, 16, 295 (1974)
303. H. H. Weetall, N. B. Havewala, W. H. Pitcher, Jr., C. C. Detar, W. P. Vann and S. Yaverbaum, *ibid.*, 16, 689 (1974)
304. W. F. Shipe, G. Senyk and H. H. Weetall, *J. Dairy Sci.*, 55, 647 (1972)
305. K. Mosbach and B. Danielsson, *Biochim. Biophys. Acta*, 364, 140 (1974)
306. H. H. Weetall and W. P. Vann, *Biotechnol. Bioeng.*, 18, 105 (1976)
307. M. Cheryan, P. J. van Wyk, N. F. Olson and T. Richardson, *ibid.*, 17, 585 (1975)
308. M. Cheryan, P. J. van Wyk, T. Richardson and N. F. Olson, *ibid.*, 18, 273 (1976)
309. R. W. Coughlin, M. Aizawa, B. F. Alexander and M. Charles, *ibid.*, 17, 515 (1975)
310. H. H. Weetall and C. C. Detar, *ibid.*, 16, 1537 (1974)
311. P. E. Market, P. F. Greenfield and J. R. Kittrell, *ibid.*, 17, 285 (1975)
312. D. B. Johnson and M. Costelloe, *ibid.*, 18, 421 (1976)
313. D. Thornton, A. Flynn and D. B. Johnson, *ibid.*, 17, 1679 (1975)
314. E. V. Leemputten and M. Horisberger, *ibid.*, 16, 385 (1974)
315. J. C. Bouin, M. T. Atallah and H. O. Hultin, *Biochim. Biophys. Acta*, 438, 23 (1976)
316. M. T. Atallah, B. P. Wasserman and H. O. Hultin, *Biotechnol. Bioeng.*, 18, 1833 (1976)
317. J. E. Brotherton, A. Emery and V. W. Rodwell, *Biotechnol. Bioeng.*, 18, 527 (1976)
318. D. D. Lee, Y. Y. Lee, P. J. Reilly, E. V. Collins, Jr. and G. T. Tsao, *ibid.*, 18, 253 (1976)
319. R. Haynes and K. A. Walsh, *Biochem. Biophys. Res. Commun.*, 36, 235 (1969)
320. D. L. Gardner, R. D. Felb, B. C. Kim and D. C. Emmerling, *Trans. Amer. Soc. Artif. Int. Organs*, 17, 239 (1971)
321. C. Horvath, *Biochim. Biophys. Acta*, 358, 164 (1974)
322. H. J. Kunz and M. Stastny, *Clin. Chem.*, 20, 1018 (1974)
323. R. E. Altomare, P. F. Greenfield and J. R. Kittrell, *Biotechnol. Bioeng.*, 16, 1675 (1974)
324. C. C. Liu, E. J. Lahoda, R. T. Galasco and L. B. Wingard, Jr., *ibid.*, 17, 1695 (1975)
325. I. Ugi, W. Betz, U. Fetzer and K. Offermann, *Chem. Ber.*, 94, 2814 (1961)
326. I. Ugi, *Angew. Chem.*, 74, 9 (1962)
327. R. Axén, P. Vretblad and J. Porath, *Acta Chem. Scand.*, 25, 1129 (1971)
328. P. Vretblad and R. Axén, *FEBS Letters*, 18, 254 (1971)
329. L. Goldstein, A. Freeman and M. Sokolovsky, *Biochem. J.*, 143, 497 (1974)

330. P. Vretblad and R. Axén, *Acta Chem. Scand.*, **27**, 2769 (1973)
331. J. Carlsson, R. Axén, K. Brocklehurst and E. M. Crook, *Eur. J. Biochem.*, **44**, 189 (1974)
332. R. Axén, H. Drevin and J. Carlsson, *Acta Chem. Scand.*, **29**, 471 (1975)
333. J. Carlsson, R. Axén and T. Unge, *Eur. J. Biochem.*, **59**, 567 (1975)
334. F. B. Weakley and C. L. Mehltretter, *Biotechnol. Bioeng.*, **15**, 1189 (1973)
335. E. Brown and A. Racois, *Tetrahedron*, **30**, 675 (1974)
336. E. Brown and R. Joyeau, *Makromol. Chem.*, **175**, 1961 (1974)
337. J. F. Kennedy and C. E. Doyle, *Carbohydrate Res.*, **28**, 89 (1973)
338. P. A. Biondi, M. Pace, O. Brenna and P. G. Pietta, *Eur. J. Biochem.*, **61**, 171 (1976)
339. J. Schnapp and Y. Shalitin, *Biochem. Biophys. Res. Commun.*, **70**, 8 (1976)
340. S. Ikeda and S. Fukui, *ibid.*, **52**, 482 (1973)
341. H. Ozawa and M. Iizuka, Abstract of Annual Meeting of the Society of Fermentation Technology, Japan, p. 74 (1969)
342. H. Ozawa, *J. Biochem.*, **62**, 419 (1967)
343. H. Ozawa, Japanese Patent 67-8910 (1967)
344. H. Ozawa, Abstract of Annual Meeting of the Society of Biochemistry, Japan p. 623 (1966)
345. H. Ozawa, *J. Biochem.*, **62**, 531 (1967)
346. G. Sodini, V. Baroncelli, M. Canella and P. Renzi, *Italian J. Biochem.*, **23**, 121 (1974)
347. B. K. Ahn, S. K. Wolfson, Jr. and S. J. Yao, *Bioelectrochem. Bioenergetics*, **2**, 142 (1975)
348. S. Avrameas, *Immunochemistry*, **6**, 43 (1969)
349. S. Avrameas and T. Ternynck, *ibid.*, **6**, 53 (1969)
350. A. Schejter and A. Bar-Eli, *Arch. Biochem. Biophys.*, **136**, 325 (1970)
351. A. K. Ferrier, T. Richardson and N. F. Olson, *Enzymologia*, **42**, 273 (1972)
352. F. A. Quiocho and F. M. Richards, *Proc. Nat. Acad. Sci.*, **52**, 833 (1964)
353. F. A. Quiocho and F. M. Richards, *Biochemistry*, **5**, 4062 (1966)
354. W. H. Bishop, F. A. Quiocho and F. M. Richards, *ibid.*, **5**, 4077 (1966)
355. S. Avrameas and B. Guilbert, *Biochimie*, **53**, 603 (1971)
336. E. F. Jansen, Y. Tomimatsu and A. C. Olson, *Arch. Biochem. Biophys.*, **144**, 394 (1971)
357. Y. Tomimatsu, E. F. Jansen, W. Gaffield and A. C. Olson, *J. Colloid Interface Sci.*, **36**, 51 (1971)
358. E. F. Jansen and A. C. Olson, *Arch. Biochem. Biophys.*, **129**, 221 (1969)
359. P. R. Witt, Jr., R. A. Sair, T. Richardson and N. F. Olson, *Brewers Dig.*, October, p. 70 (1970)
360. M. Ottesen and B. Svensson, *Compt. Rend. Trav. Lab. Carlsberg*, **38**, 171 (1971)
361. K. Ogata, M. Ottesen and I. Svendsen, *Biochim. Biophys. Acta*, **159**, 403 (1968)
362. E. K. Bouman and L. H. Goodson, *Anal. Chem.*, **37**, 1378 (1965)
363. G. G. Guilbault and D. N. Kramer, *ibid.*, **37**, 1675 (1965)
364. L. H. Goodson and W. B. Jacobs, *Anal. Biochem.*, **51**, 362 (1973)
365. S. Kasuga, K. Oike, T. Kobayashi and S. Suzuki, *Hakko Kyokaishi* (Japanese), **28**, 299 (1970)
366. T. Wieland, H. Determann and K. Bünnig, *Z. Naturforsch.*, **216**, 1003 (1966)
367. G. P. Hicks and S. J. Updike, *Anal. Chem.*, **38**, 726 (1966)
368. S. J. Updike and G. P. Hicks, *Nature*, **214**, 986 (1967)
369. L. B. Wingard, Jr., C. C. Liu and N. L. Nagda, *Biotechnol. Bioeng.*, **13**, 629 (1971)
370. S. Gestrelius, B. Mattiasson and K. Mosbach, *Eur. J. Biochem.*, **36**, 89 (1973)
371. K. Kawashima and K. Umeda, *Biotechnol. Bioeng.*, **16**, 609 (1974)
372. G. G. Guilbault and E. Hrabankova, *Anal. Chem.*, **42**, 1779 (1970)
373. H. D. Brown, A. B. Patel and S. K. Chattopadhyay, *J. Chromatog.*, **35**, 103 (1968)

374. P. Bernfeld and J. Wan, *Science*, **142**, 678 (1963)
375. Y. Degani and T. Miron, *Biochim. Biophys. Acta*, **212**, 362 (1970)
376. T. T. Ngo, P. S. Bunting and K. J. Laidler, *Can. J. Biochem.*, **53**, 11 (1975)
377. H. M. Walton, J. E. Eastman and A. E. Staley, *Biotechnol. Bioeng.*, **15**, 951 (1973)
378. C. Gruesbeck and H. F. Rase, *Ind. Eng. Chem. Prod. Res. Develop.*, **11**, 74 (1972)
379. H. Maeda, H. Suzuki and A. Yamauchi, *Biotechnol. Bioeng.*, **15**, 607 (1973)
380. H. Maeda, A. Yamaguchi and H. Suzuki, *Biochim. Biophys. Acta*, **315**, 18 (1973)
381. R. A. Llenado and G. A. Rechnitz, *Anal. Chem.*, **43**, 1457 (1971)
382. P. S. Bunting and K. J. Laidler, *Biochemistry*, **11**, 4477 (1972)
383. A. Dahlqvist, B. Mattiasson and K. Mosbach, *Biotechnol. Bioeng.*, **15**, 395 (1973)
384. S. Usami and Y. Kuratsu, *Hakkokogaku Zasshi* (Japanese), **51**, 789 (1973)
385. H. D. Brown, A. B. Patel and S. K. Chattopadhyay, *J. Biomed. Mater, Res.*, **2**, 231 (1968)
386. K. Mosbach and R. Mosbach, *Acta Chem. Scand.*, **20**, 2807 (1966)
387. J. Dobó, *Acta Chim. Acad. Sci. Hung.*, **63**, 453 (1970)
388. A. Johansson, J. Lundberg, B. Mattiasson and K. Mosbach, *Biochim. Biophys. Acta*, **304**, 217 (1973)
389. G. J. Bartling, H. D. Brown and S. K. Chattopadhyay, *Nature*, **243**, 342 (1973)
390. T. Mori, T. Tosa and I. Chibata, *Cancer Res.*, **34**, 3066 (1974)
391. T. Mori, T. Sato, T. Tosa and I. Chibata, *Enzymologia*, **43**, 213 (1972)
392. G. J. Papariello, A. K. Mukherji and C. M. Shearer, *Anal. Chem.*, **45**, 790 (1973)
393. P. Bernfeld, R. E. Bieber and P. C. Macdonnell, *Arch. Biochem. Biophys.*, **127**, 779 (1968)
394. P. Bernfeld and R. E. Bieber, *ibid.*, **131**, 587 (1969)
395. G. W. Strandberg and K. L. Smiley, *Appl. Microbiol.*, **21**, 588 (1971)
396. T. Kasumi, K. Kawashima and N. Tsumura, *J. Ferment. Technol.*, **52**, 321 (1974)
397. H. Maeda, H. Suzuki, A. Yamauchi and A. Sakimae, *Biotechnol. Bioeng.*, **17**, 119 (1975)
398. S. Fukui, A. Tanaka, T. Iida and E. Hasegawa, *FEBS Letters*, **66**, 179 (1976)
399. H. Maeda, H. Suzuki, A. Yamauchi and A. Sakimae, *Biotechnol. Bioeng.*, **16**, 1517 (1974)
400. H. Maeda, *ibid.*, **17**, 1571 (1975)
401. S. N. Pennington, H. D. Brown, A. B. Patel and C. O. Knowles, *Biochim. Biophys. Acta*, **167**, 479 (1968)
402. J. C. W. Østergaard and S. C. Martiny, *Biotechnol. Bioeng.*, **15**, 561 (1973)
403. T. M. S. Chang, *Proc. Can. Fed. Biol. Soc.*, **12**, 62 (1969)
404. T. M. S. Chang, *Nature*, **229**, 117 (1971)
405. T. Mori, T. Sato, Y. Matuo, T. Tosa and I. Chibata, *Biotechnol. Bioeng.*, **14**, 663 (1972)
406. T. Mori, T. Tosa and I. Chibata, *Biochim. Biophys. Acta*, **321**, 653 (1973)
407. S. N. Levine and W. C. Lacourse, *J. Biomed. Mater, Res.*, **1**, 275 (1967)
408. T. M. S. Chang, *Trans. Amer. Soc. Artif. Int. Organs*, **12**, 13 (1966)
409. R. E. Sparks, O. Lindan and M. H. Litt, Annual Progress Report, Chemical Engineering Science Division, Case Western Reserve Univ., Cleveland, Ohio 1968
410. T. M. S. Chang and F. C. MacIntosh, *Pharmacologist*, **6**, 198 (1964)
411. T. M. S. Chang, F. C. MacIntosh and S. G. Mason, *Can. J. Physiol. Pharmacol.*, **44**, 115 (1966)
412. R. C. Boguslaski and A. M. Janik, *Biochim. Biophys. Acta*, **250**, 266 (1971)
413. M. Kitajima, S. Miyano and A. Kondo, *Kogyokagaku Zasshi* (Japanese), **72**, 493 (1969)
414. M. Kitajima and A. Kondo, *Bull. Chem. Soc. Japan*, **44**, 3201 (1971)
415. D. L. Gardner and D. Emmerling, *Chem. Eng. News*, **49**, 31 (1971)
416. D. T. Wadiak and R. G. Carbonell, *Biotechnol. Bioeng.*, **17**, 1157 (1975)
417. T. M. S. Chang, *Enzyme*, **14**, 95 (1973)

418. J. Campbell and T. M. S. Chang, *Biochem. Biophys. Res. Commun.*, **69**, 562 (1976)
419. T. M. S. Chang, *Science*, **146**, 524 (1964)
420. T. M. S. Chang and M. J. Poznansky, *Nature*, **218**, 243 (1968)
421. A. D. Mogensen and W. R. Vieth, *Biotechnol. Bioeng.*, **15**, 467 (1973)
422. M. J. Poznansky and T. M. S. Chang, *Proc. Can. Fed. Biol. Soc.*, **12**, 54 (1969)
423. T. M. S. Chang, *Biochem. Biophys. Res. Commun.*, **44**, 1531 (1971)
424. J. Campbell and T. M. S. Chang, *Biochim. Biophys. Acta*, **397**, 101 (1975)
425. M. A. Paine and R. G. Carbonell, *Biotechnol. Bioeng.*, **17**, 617 (1975)
426. G. Gregoriadis, P. D. Leathwood and B. E. Ryman, *FEBS Letters*, **14**, 95 (1971)
427. S. W. May and N. N. Li, *Biochem. Biophys. Res. Commun.*, **47**, 1179 (1972)
428. G. Sessa and G. Weissmann, *J. Biol. Chem.*, **245**, 3295 (1970)
429. G. Dapergolas, E. D. Neerunjun and G. Gregoriadis, *FEBS Letters*, **63**, 235 (1976)
430. L. D. Steger and R. J. Desnick, *Biochim. Biophys. Acta*, **464**, 530 (1977)
431. J. Kirimura and J. Yoshida, Japanese Patent 64-27492 (1964)
432. M. Pecht and Y. Levin, *Biochem. Biophys. Res. Commun.*, **46**, 2054 (1972)
433. H. Negoro, *J. Ferment. Technol.*, **50**, 136 (1972)
434. S. J. Updike, R. T. Wakamiya and E. N. Lightfoot, Jr., *Science*, **193**, 681 (1976)
435. A. Emery, J. Sorenson, M. Kolarik, S. Swanson and H. Lim, *Biotechnol. Bioeng.*, **16**, 1359 (1974)
436. M. D. Lilly, *ibid.*, **13**, 589 (1971)
437. H. H. Weetall and N. Weliky, *Anal. Biochem.*, **14**, 160 (1966)
438. H. P. Gregor and P. W. Rauf, *Biotechnol. Bioeng.*, **17**, 445 (1975)
439. R. Goldman, H. I. Silman, S. R. Caplan, O. Kedem and E. Katchalski, *Science*, **150**, 758 (1965)
440. R. Goldman, O. Kedem, I. H. Silman, S. R. Caplan and E. Katchalski, *Biochemistry*, **7**, 486 (1968)
441. R. Goldman, O. Kedem and E. Katchalski, *ibid.*, **7**, 4518 (1968)
442. F. Leuschner, Ger. Patent, 1227855 (1966)
443. G. Broun, C. Tran-Minh, D. Thomas, D. Domurado and E. Selegny, *Trans. Amer. Soc. Artif. Int. Organs*, **17**, 341 (1971)
444. K. Venkatasubramanian, W. R. Vieth and S. S. Wang, *J. Ferment. Technol.*, **50**, 600 (1972)
445. P. R. Coulet, J. H. Julliard and D. C. Gautheron, *Biotechnol. Bioeng.*, **16**, 1055 (1974)
446. J. H. Julliard, C. Godinot and D. C. Gautheron, *FEBS Letters*, **14**, 185 (1971)
447. P. R. Coulet, C. Godinot and D. C. Gautheron, *Biochim. Biophys. Acta*, **391**, 272 (1975)
448. K. Venkatasubramanian, R. Saini and W. R. Vieth, *J. Food Sci.*, **40**, 109 (1975)
449. H. D. Chu, J. G. Leeder and S. G. Gilbert, *ibid.*, **40**, 641 (1975)
450. M. Kubo, I. Karube and S. Suzuki, *Biochem. Biophys. Res. Commun.*, **69**, 731 (1976)
451. J. Jakubowski, J. R. Giacin, D. H. Kleyn, S. G. Gilbert and J. G. Leeder, *J. Food Sci.*, **40**, 467 (1975)
452. Y. Nakamoto, I. Karube and S. Suzuki, *J. Ferment. Technol.*, **53**, 595 (1975)
453. Y. Nakamoto, I. Karube and S. Suzuki, *Biotechnol. Bioeng.*, **17**, 1387 (1975)
454. I. Karube, Y. Nakamoto, K. Namba and S. Suzuki, *Biochim. Biophys. Acta*, **429**, 975 (1976)
455. K. Venkatasubramanian and W. R. Vieth, *Biotechnol. Bioeng.*, **15**, 583 (1973)
456. I. Karube and S. Suzuki, *Biochem. Biophys. Res. Commun.*, **47**, 51 (1972)
457. J. G. Dillon and C. W. R. Wade, *Biotechnol. Bioeng.*, **18**, 133 (1976)
458. Y. Inada, S. Hirose, M. Okada and H. Mihama, *Enzyme*, **20**, 188 (1975)
459. J. P. Bollmeier and S. Middleman, *Biotechnol. Bioeng.*, **16**, 859 (1974)
460. W. J. Blaedel, T. R. Kissel and R. C. Boguslaski, *Anal. Chem.*, **44**, 2030 (1972)
461. W. E. Hornby and H. Filippusson, *Biochim. Biophys. Acta*, **220**, 343 (1970)

462. W. E. Hornby, D. J. Inman and A. McDonald, *FEBS Letters*, **23**, 114 (1972)
463. D. J. Inman and W. E. Hornby, *Biochem. J.*, **129**, 255 (1972)
464. D. J. Inman and W. E. Hornby, *ibid.*, **137**, 25 (1974)
465. J. Campbell, W. E. Hornby and D. L. Morris, *Biochim. Biophys. Acta*, **384**, 307 (1975)
466. T. T. Ngo, D. Narinesingh and K. J. Laidler, *Biotechnol. Bioeng.*, **18**, 119 (1976)
467. G. F. Senyk, E. C. Lee, W. F. Shipe, L. F. Hood and T. W. Downes, *J. Food Sci.*, **40**, 288 (1975)
468. T. T. Ngo, *Can. J. Biochem.*, **54**, 62 (1976)
469. T. Mori, R. Sano, Y. Iwasawa, T. Tosa and I. Chibata, *J. Solid-Phase Biochem.*, **1**, 15 (1976)
470. D. Dinelli, *Process Biochem.*, **7**, 9 (1972)
471. W. Marconi, S. Gulinelli and Morisi, *Biotechnol. Bioeng.*, **16**, 501 (1974)
472. S. Glovenco, F. Morisi and P. Pansolli, *FEBS Letters*, **36**, 57 (1973)
473. T. A. Butterworth, D. I. C. Wang and A. J. Sinskey, *Biotechnol. Bioeng.*, **12**, 615 (1970)
474. J. R. Wykes, P. Dunnill and M. D. Lilly, *Biochim. Biophys. Acta*, **250**, 522 (1971)
475. J. J. Marshall and W. J. Whelan, *Chem. Ind.* (*London*), No. 25, 701 (1971)
476. D. I. C. Wang and A. E. Humphrey, *Chem. Eng.*, **76**, 108 (1969)
477. T. K. Ghose and J. A. Kostick, *Biotechnol. Bioeng.*, **12**, 921 (1970)
478. S. P. O'Neill, J. R. Wykes, P. Dunnill and M. D. Lilly, *ibid.* **13**, 319 (1971)
479. J. C. Davis, *ibid.*, **16**, 1113 (1974)
480. D. J. Fink and V. W. Rodwell, *ibid.*, **17**, 1029 (1975)
481. K. L. Smiley, D. E. Hensley and H. J. Gasdorf, *Appl. Environ. Microbial.*, **31**, 615 (1976)
482. H. D. Brown, S. K. Chattopadhyay and A. B. Patel, *Biochem. Biophys. Res. Commun.*, **25**, 304 (1966)
483. K. Ohmiya, C. Terao, S. Shimizu and T. Kobayashi, *Agr. Biol. Chem.*, **39**, 491 (1975)
484. J. C. Brown, H. E. Swaisgood and H. R. Horton, *Biochem. Biophys. Res. Commun.*, **48**, 1068 (1972)
485. Y. Ishimatsu, S. Shigesada and S. Kimura, Japanese Patent Kokai, 76-44688 (1976)
486. T. Hattori and C. Furusaka, *J. Biochem.*, **50**, 312 (1961)
487. T. Hattori and C. Furusaka, *ibid.*, **48**, 831 (1960)
488. T. Fujii, K. Hanamitsu, R. Izumi, T. Yamada and T. Watanabe, Japanese Patent Kokai, 73-99393 (1973)
489. T. Fujii, K. Matsumoto, Y. Shibuya, K. Hanamitsu, T. Yamaguchi and T. Watanabe, British Patent, 1,347,665 (1974)
490. D. E. Johnson and A. Ciegler, *Arch. Biochem. Biophys.*, **130**, 384 (1969)
491. G. Corrieu, H. Blachere, A. Ramirez, J. M. Navarro, G. Durand, I. N. S. A. Toulouse, B. Duteurtre and M. Moll, Fifth International Fermentation Symposium, Berlin, (1976), p. 294
492. Novo Industri, Japanese Patent Kokai, 76-51580 (1976)
493. I. Chibata, T. Tosa and T. Sato, *Appl. Microbiol.*, **27**, 878 (1974)
494. H. Nishimaru, C. Izumi, S. Narita and K. Yamada, Japanese Patent Kokai, 75-140680 (1975)
495. I. Chibata, J. Kato, T. Watanabe and T. Uchida, Japanese Patent Kokai, 75-135290 (1975)
496. S. Kinoshita, M. Muranaka and H. Okada, *J. Ferment. Technol.*, **53**, 223 (1975)
497. K. Yamamoto, T. Sato, T. Tosa and I. Chibata, *Biotechnol. Bioeng.*, **16**, 1601 (1974)
498. S. Nonomura, M. Watanabe and T. Inagaki, Japanese Patent Kokai, 76-48488 (1976)
499. S. Yagi, Y. Toda and T. Minoda, Annual Meeting of the Agricultural Chemical

Society of Japan, at Kyoto, April 4 (1976), p. 414
500. K. Yamamoto, T. Tosa, K. Yamashita and I. Chibata, *Eur. J. Appl. Microbiol.*, 3, 169 (1976)
501. S. Shimizu, H. Morioka, Y. Tani and K. Ogata, *J. Ferment. Technol.*, 53, 77 (1975)
502. I. Karube, K. Matsunaga, S. Tsuru and S. Suzuki, *Biochim. Biophys. Acta*, 444, 338 (1976)
503. W. Slowinski and S. E. Charm, *Biotechnol. Bioeng.*, 15, 973 (1973)
504. T. Sato, T. Tosa and I. Chibata, *Eur. J. Appl. Microbiol.*, 2, 153 (1976)
505. I. Chibata, T. Kakimoto and K. Nabe, Japanese Patent Kokai, 74-81591 (1974)
506. I. Chibata, T. Kakimoto and K. Nabe, *ibid.*, 74-81590 (1974)
507. S. R. Saif, Y. Tani and K. Ogata, *J. Ferment. Technol.*, 53, 380 (1975)
508. C. K. A. Martin and D. Perlman, *Biotechnol. Bioeng.*, 18, 217 (1976)
509. O. Kanemitsu, Japanese Patent Kokai, 75-132181 (1975)
510. I. Chibata, T. Tosa, T. Sato and K. Yamamoto, Japanese Patent Kokai, 75-100289 (1975)
511. K. Yamamoto, T. Sato, T. Tosa and I. Chibata, *Biotechnol. Bioeng.*, 16, 1589 (1974)
512. N. E. Franks, *Biochim. Biophys. Acta*, 252, 246 (1971)
513. I. Chibata, T. Tosa and T. Sato, Japanese Patent Kokai, 74-132290 (1974)
514. U. Hackel, J. Klein, R. Megnet and F. Wagner, *Eur. J. Appl. Microbiol.*, 1, 291 (1975)
515. K. Mosbach and P. O. Larsson, *Biotechnol. Bioeng.*, 12, 19 (1970)
516. K. Mosbach and R. Mosbach, *Acta Chem. Scand.*, 20, 2807 (1966)
517. S. J. Updike, D. R. Harris and E. Shrago, *Nature*, 224, 1122 (1969)
518. N. Miyata and T. Kikuchi, Japanese Patent Kokai, 76-133484 (1976)
519. K. Toda and M. Shoda, *Biotechnol. Bioeng.*, 17, 481 (1975)
520. I. Chibata, T. Tosa and T. Mori, Japanese Patent Kokai, 74-30582 (1974)
521. W. Marconi, F. Bartoli, F. Cerere, G. Galli and F. Morisi, *Agr. Biol. Chem.*, 39, 277 (1975)
522. SNAM Progetti, Belgian Patent, 782,646 (1972)
523. M. J. Kolarik, B. J. Chen, A. H. Emery, Jr. and H. C. Lim, *Immobilized Enzymes in Food and Microbial Process* (ed. A. C. Olson and C. L. Cooney), p. 71, Plenum Press, 1974
524. A. M. Brownstein, W. R. Vieth and A. Constantinides, Symposium on "Advance in Immobilized Enzyme Systems", American Chemical Society Meeting, at Atlantic City, September 9 (1974)
525. I. Karube, S. Mitsuda and S. Suzuki, Annual Meeting of the Society of Fermentation Technology, Japan, at Osaka, October 25, p. 127 (1976)
526. W. R. Vieth, S. S. Wang and R. Saini, *Biotechnol. Bioeng.*, 15, 565 (1973)
527. S. S. Wang, W. R. Vieth and A. Constantinides, *Enzyme Engineering*, 2, 123 (1973)
528. K. Venkatasubramanian, R. Saini and W. R. Vieth, *J. Ferment. Technol.*, 52, 268 (1974)
529. H. Yamada, M. Yamada, E. Nagasawa, H. Kumagai, T. Hino and S. Okamura, Annual Meeting of the Agricultural Chemical Society of Japan, at Sapporo, July 24, p. 336, (1975)
530. H. Kumagai, S. Sezima, H. Yamada, T. Hino and S. Okamura, *ibid.*, at Kyoto, April 2, p. 233, (1976)
531. K. Hirano, I. Karube and S. Suzuki, *Biotechnol. Bioeng.*, 19, 311 (1977)
532. S. Yuta, R. R. Bhatt, T. Yoshida and K. Taguchi, Annual Meeting of the Society of Fermentation Technology, Japan, at Osaka, October 30, p. 250, (1975)
533. N. Tsumura, T. Kasumi and M. Ishikawa, *Rept. Nat. Food. Res. Inst.* (Japanese), No. 31, p. 71 (1976)
534. S. Tsumura and T. Kasumi, Japanese Patent Kokai, 77-44285 (1977)
535. A. Nanba and Y. Matuo, Annual Meeting of the Agricultural Chemical Society of Japan, at Fukuoka, April 4, p. 251, (1970)

536. Y. Takasaki and A. Kanbayashi, *Kogyo Gijutsuin Biseibutsu Kogyo Gijutsu Kenkyusho Hokoku* (Japanese), No. 37, 31 (1969)
537. Y. Kosugi and H. Suzuki, *J. Ferment. Technol.*, **51**, 895 (1973)
538. N. Tsumura, Annual Meeting of the Society of Fermentation Technology, Japan, at Osaka, November 14, p. 81, (1969)
539. R. R. Mohan and N. N. Li, *Biotechnol. Bioeng.*, **17**, 1137 (1975)
540. C. R. Lowe and P. D. G. Dean, *FEBS Letters*, **14**, 313 (1971)
541. C. R. Lowe and P. D. G. Dean, *ibid.*, **18**, 31 (1971)
542. S. Barry and P. O'Carra, *Biochem. J.*, **135**, 595 (1973)
543. D. B. Craven, M. J. Harvey and P. D. G. Dean, *FEBS Letters*, **38**, 320 (1974)
544. M. Lindberg, P. O. Larsson and K. Mosbach, *Eur. J. Biochem.*, **40**, 187 (1973)
545. C. R. Lowe, M. J. Harvey and P. D. G. Dean, *ibid.*, **41**, 347 (1974)
546. H. J. Comer, D. B. Craven, M. J. Marvey, A. Atkinson and P. D. G. Dean, *ibid.*, **55**, 201 (1975)
547. K. Mosbach, *Methods in Enzymology*, vol. 34, p. 229, Academic Press, 1974
548. A. K. Grover and G. G. Hammes, *Biochim. Biophys. Acta*, **356**, 309 (1974)
549. C. R. Lowe, K. Masbach and P. D. G. Dean, *Biochem. Biophys. Res. Commun.*, **48**, 1004 (1972)
550. M. J. Harvey, D. B. Craven, C. R. Lowe and P. D. G. Dean, *Methods in Enzymology*, vol. 34, p. 242, Academic Press, 1974
551. K. Mosbach, H. Guilford, R. Ohlsson and M. Scott, *Biochem. J.*, **127**, 625 (1972)
552. C. R. Lowe, M. J. Harvey, D. B. Craven and P. D. G. Dean, *ibid.*, **133**, 499 (1973)
553. C. R. Lowe, M. J. Harvey, D. B. Craven, M. A. Kerfoot, M. E. Hollows and P. D. G. Dean, *ibid.*, **133**, 507 (1973)
554. S. Barry and P. O'Carra, *FEBS Letters*, **37**, 134 (1973)
555. P. O. Larsson and K. Mosbach, *Biotechnol. Bioeng.*, **13**, 393 (1971)
556. C. R. Lowe and P. D. G. Dean, *Biochem. J.*, **133**, 515 (1973)
557. B. Tabahoff and J. P. von Wartburg, *Biochem. Biophys. Res. Commun.*, **63**, 957 (1975)
558. N. O. Kaplan, J. Everse, J. E. Dixon, F. E. Stolzenbach, C.-Y. Lee, C.-L. Lee, S. S. Taylor and K. Mosbach, *Proc. Nat. Acad. Sci.*, **71**, 3450 (1970)
559. J. D. Hocking and J. I. Harris, *FEBS Letters*, **34**, 280 (1973)
560. R. Lamed, Y. Levin and M. Wilchek, *Biochim. Biophys. Acta*, **304**, 231 (1973)
561. P. T. Gilham, *Methods in Enzymology*, vol. 21, p. 191, Academic Press, 1971
562. K. Yagi, *Seikagaku Jikken Koza* (Japanese), vol. 15, p. 31, Tokyo Kagaku Dozin, 1975
563. M. Wilchek and R. Lamed, *Methods in Enzymology*, vol. 34, p. 475, Academic Press, 1974
564. R. Lamed, Y. Levin and A. Oplatka, *Biochim. Biophys. Acta*, **305**, 163 (1973)
565. R. Lamed and A. Oplatka, *Biochemistry*, **13**, 3137 (1974)
566. K. Mosbach, P. O. Larsson, P. Brodelius, H. Guilford and M. Lindberg, *Enzyme Engineering*, **2**, 237 (1974)
567. M. K. Weibel, H. H. Weetall and H. J. Bright, *Biochem. Biophys. Res. Commun.*, **44**, 347 (1971)
568. J. R. Wykes, P. Dunnill and M. D. Lilly, *Biochim. Biophys. Acta*, **286**, 260 (1972)
569. R. Ohlsson, P. Brodelius and K. Mosbach, *FEBS Letters*, **25**, 234 (1974)
570. D. B. Craven, H. J. Harvey, C. R. Lowe and P. D. G. Dean, *Eur. J. Biochem.*, **41**, 329 (1974)
571. M. J. Harvey, C. R. Lowe, D. B. Craven and P. D. G. Dean, *ibid.*, **41**, 335 (1974)
572. C. R. Lowe, M. J. Harvey and P. D. G. Dean, *ibid.*, **41**, 347 (1974)
573. M. J. Harvey, C. R. Lowe and P. D. G. Dean, *ibid.*, **41**, 353 (1974)
574. L. Anderson, H. Jörnvall, Å. Åkeson and K. Mosbach, *Biochim. Biophys. Acta*, **364**, 1 (1974)
575. P. Brodelius, P. O. Larsson and K. Mosbach, *Eur. J. Biochem.*, **47**, 81 (1974)

576. C. R. Lowe, M. J. Harvey and P. D. G. Dean, *ibid.*, **41**, 341 (1974)
577. I. P. Trayer and H. R. Trayer, *Biochem. J.*, **141**, 775 (1974)
578. P. Brodelius and K. Mosbach, *FEBS Letters*, **35**, 223 (1973)
579. C. R. Lowe and K. Mosbach, *Eur. J. Biochem.*, **52**, 99 (1975)
580. I. P. Trayer, H. R. Trayer, D. A. Small and R. C. Bottomley, *Biochem. J.*, **139**, 609 (1974)
581. R. Barker, I. P. Trayer and R. L. Hill, *Methods in Enzymology*, vol. 34, p. 479, Academic Press, 1974
582. M. Lindberg and K. Mosbach, *Eur. J. Biochem.*, **53**, 481 (1975)
583. C. Y. Lee, D. A. Lappi, B. Wermuth, J. Everse and N. O. Kaplan, *Arch. Biochem. Biophys.*, **163**, 561 (1974)
584. M. Wilchek, Y. Salomon, M. Lowe and Z. Selinger, *Biochem. Biophys. Res. Commun.*, **45**, 1177 (1971)
585. B. Jergil and K. Mosbach, *Method in Enzymology*, vol. 34, p. 261, Academic Press 1974
586. G. I. Tesser, H.-U. Fish and R. Schwyzer, *FEBS Letters*, **23**, 56 (1972)
587. O. Berglund and F. Eckstein, *Eur. J. Biochem.*, **28**, 492 (1972)
588. O. Berglund and F. Eckstein, *Methods in Enzymology*, vol. 34, p. 253, Academic Press, 1974
589. O. Berglund, *J. Biol. Chem.*, **247**, 7270 (1972)
590. L. Thelander, *ibid.*, **248**, 4591 (1973)
591. J. A. Fuchs, H. O. Karlström, H. R. Warner and P. Reichard, *Nature New Biol.*, **238**, 69 (1972)
592. K. Olesen, E. Hippe and E. Hober, *Biochim. Biophys. Acta*, **243**, 66 (1971)
593. C. Arsenis and D. B. McCormick, *J. Biol. Chem.*, **241**, 330 (1966)
594. M. N. Kazarinobb, C. Arsenis and D. B. McCormick, *Methods in Enzymology*, vol. 34, p. 300, Academic Press, 1974
595. T. Imai and J. Tobari, *Seikagaku* (Japanese), **46**, 508 (1974)
596. G. Blankenhorn, D. T. Osuga, H. S. Lee and R. Feeney, *Biochim. Biophys. Acta*, **386**, 470 (1975)
597. C. Arsenis and D. B. McCormick, *J. Biol. Chem.*, **239**, 3093 (1964)
598. J. V. Miller, Jr., P. Cuatrecasas and E. B. Thompson, *Biochim. Biophys. Acta*, **276**, 407 (1972)
599. J. V. Miller, Jr., P. Cuatrecasas and E. B. Thompson, *Proc. Nat. Acad. Sci.*, **68**, 1014 (1971)
600. E. B. Thompson, *Methods in Enzymology*, vol. 34, p. 294, Academic Press, 1974
601. R. Collier and G. Kohlhaw, *Anal. Biochem.*, **42**, 48 (1971)
602. E. Ryan and P. F. Fottrell, *FEBS Letters*, **23**, 73 (1972)
603. S. Ikeda, H. Hara and S. Fukui, *Biochim. Biophys. Acta*, **372**, 400 (1974)
604. S. Ikeda, H. Hara, S. Sugimoto and S. Fukui, *FEBS Letters*, **56**, 307 (1975)
605. S. Fukui, S. Ikeda, M. Fujimura, H. Yamada and H. Kumagai, *Eur. J. Biochem.*, **51**, 155 (1975)
606. H. Hara, S. Sugimoto, S. Ikeda and S. Fukui, *Seikagaku* (Japanese), **46**, 655 (1974)
607. R. Yamada and H. P. C. Hogenkamp, *J. Biol. Chem.*, **247**, 6266 (1972)
608. R. H. Allen and P. W. Majerus, *ibid.*, **247**, 7695, 7702, 7709 (1972)
609. R. H. Allen and C. S. Meklman, *ibid.*, **248**, 3660, 3670 (1973)
610. E. L. Lien, L. Ellenbogen, P. Y. Law and J. M. Wood, *Biochem. Bilphys. Res. Commun.*, **55**, 730 (1973)
611. E. L. Lien, L. Ellenbogen, P. Y. Law and J. M. Wood, *J. Biol. Chem.*, **249**, 890 (1974)
612. P. V. Danenberg and R. J. Langenbach, *Biochem. Biophys. Res. Commun.*, **49**, 1029 (1972)
613. P. V. Danenberg and C. Heidelberger, *Methods in Enzymology*, vol. 34, p. 520, Academic Press, 1974
614. R. Barker, K. W. Olsen, J. H. Shaper and R. L. Hill, *J. Biol. Chem.*, **247**, 7135

(1972)
615. R. K. Wierenge, J. D. Huizinga, W. Gaastra, G. W. Welling and J. J. Beintema, *FEBS Letters*, **31**, 181 (1973)
616. R. J. Jackson, R. M. Wolcott and T. Shiota, *Biochem. Biophys. Res. Commun.*, **51**, 428 (1973)
617. C. Godinot, J. H. Julliard and D. C. Gautheron, *Anal. Biochem.*, **61**, 264 (1974)
618. D. L. Robberson and N. Davidson, *Biochemistry*, **11**, 533 (1972)
619. P. Cuatrecasas and M. Wilchek, *Biochem. Biophys. Res. Commun.*, **33**, 235 (1974)
620. Y. Matsuo, T. Tosa and I. Chibata, *Biochim. Biophys. Acta*, **338**, 520 (1974)
621. A. Matsuura, A. Iwashima and Y. Nose, *Biochem. Biophys. Res. Commun.*, **51**, 241 (1973)
622. J. S. Erickson and C. K. Mathews, *ibid.*, **43**, 1164 (1971)
623. W. H. Scouten, F. Torok and W. Gitomer, *Biochim. Biophys. Acta*, **309**, 521 (1973)
624. T. Tosa, T. Sato, T. Mori, K. Yamamoto, I. Takata, Y. Nishida and I. Chibata, *Biotechnol. Bioeng.*, **21**, 133 (1979)
625. T. Tosa, T. Sato, K. Yamamoto, I. Takata, Y. Nishida and I. Chibata, 26th International Congress of Pure and Applied Chemistry, at Tokyo, September 4, p. 267, 1977

CHAPTER 3

Properties of immobilized enzymes and microbial cells

Information on changes of enzymatic properties caused by the immobilization of enzymes and microbial cells is useful not only for the application of immobilized systems but also for the elucidation of structure-function relationships and mechanisms of enzyme reaction.

The changes of enzymatic properties are considered to be caused by the following two factors. One is changes of the enzyme itself, and the other is physical and chemical properties of the carriers used for immobilization. The former involves modification of amino acid residues in the active center of the enzyme, conformational changes of the enzyme protein, and changes in the charge on the enzyme, while the latter involves the formation of diffusion layers around immobilized enzymes, steric hindrance effects, and electrostatic interaction between carrier and substrate. The observed changes of enzymatic properties on immobilization are the result of complicated interactions of these factors, and it is difficult to determine the exact effect of a given factor in terms of changes of enzymatic properties.

3.1 SUBSTRATE SPECIFICITY

As already mentioned, when enzymes or microbial cells are immobilized, the enzyme activities often decrease, and the substrate specificities sometimes change. In particular, when enzymes acting on substrates of high molecular weight, e.g., proteolytic enzymes and amylases, are immobilized by carrier-binding methods such as peptide binding or diazo binding, significant changes of substrate specificity are generally observed, as shown in Table 3.1.

For example, trypsin immobilized by peptide binding using CM-cellulose azide showed the same activity as the native enzyme toward a synthetic substrate of low molecular weight, benzoylarginine-*p*-nitroanilide (BApNA), but showed only 3% activity compared with that of the native

108

TABLE 3.1 Substrate specificities of immobilized enzymes

Enzyme	Immobilization method	High molecular substrate	Low molecular substrate	Reference
		Relative activity†		
Glucoamylase	peptide binding (CM-cellulose azide)	15–17 amylose MW: 5×10^5	77 amylose MW: 8×10^3	1
Chymotrypsin	peptide binding (CM-cellulose azide)	1.6–2.8 (casein)	19–39 (BTpNA)	2
	Ugi reaction (lysine–Sepharose)	20 (casein)	50 (ATEE)	3
	(CM-Sephadex)	3 (casein)	15 (ATEE)	3
Trypsin	ionic binding (DEAE–Sephadex)	68 (casein)	30–100 (TAME)	4
	peptide binding (CM-cellulose azide)	1.5–3.0 (casein)	44–64 (BAEE)	2
		8–15 (clupeine)	94–108 (BApNA)	2
	(CNBr-activated Sepharose)	74 (casein)	20–53 (TAME)	4
	(isothiocyanate-activated porous glass)	45 (casein)	17–38 (TAME)	4
	alkylation (bromoacetyl cellulose)	0.23 (casein)	0.6–0.7 (LEE)	5
	carrier cross-linking (silica powder + glutaraldehyde)	15 (casein)	80 (BAEE)	6
Elastase	peptide binding (CNBr-activated Sepharose)	27 (casein)	37 (Ac–Ala–Ala–Ala–ME)	7
Papain	diazo binding (amino acid copolymer)	30 (casein)	70 (BAEE)	8
	cross-linking (glutaraldehyde)	12 (casein)	48 (BAEE)	9
		16 (hemoglobin)		9
Ficin	peptide binding (CM-cellulose azide)	4–5 (casein)	8–12 (BAEE)	10
Pronase	diazo binding (porous glass)	57 (albumin)	67 (LpNA)	11

† The activity of native enzyme was taken as 100.

enzyme toward a substrate of high molecular weight, casein.[2] Gluco-amylase immobilized by the same method had 77% activity compared with that of the native enzyme toward amylose of molecular weight 8000, but had only 15~17% of the native enzyme activity toward amylose of molecular weight over 500,000.[1]

That is, when an enzyme is immobilized by using a water-insoluble polymer carrier, the enzyme activity toward substrates of high molecular weight is markedly reduced due to steric hindrance, which obstructs the approach of the substrate to the enzyme molecule. However, the activity of immobilized enzyme toward substrates of low molecular weight is similar to that of the native enzyme because the substrate can easily approach the enzyme molecule.

Activities of native and immobilized pronase prepared by diazo binding using leucine and p-aminophenylalanine copolymer were compared toward substrates of various molecular weights.[12] As shown in Fig. 3.1, the activity of immobilized enzyme relative to that of the native enzyme decreases with increase of molecular weight of the substrate. This indicates that the reduction in the activity of immobilized enzyme acting on substrates of high molecular weight is caused by steric hindrance between the carrier and substrate. In the case of immobilized enzyme

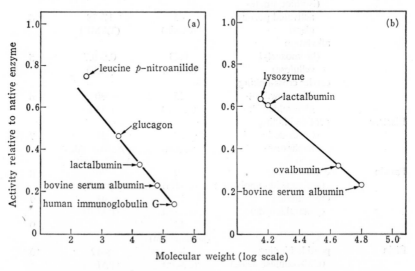

Molecular weight (log scale)

Fig. 3.1   Relationship between the activity of immobilized enzyme and the molecular weight of the substrate.[12]
(a) Immobilized pronase I. (b) Immobilized pronase II. Immobilized pronase I is different from immobilized pronase II in the amount of pronase bound to the carrier.

prepared by the entrapping method, the use of a semipermeable polymer makes it intrinsically difficult to act on a substrate of high molecular weight. An opposite phenomenon was observed in trypsin immobilized by using CNBr-activated Sepharose or isothiocyanate-activated porous glass.[4] That is, high molecular weight casein was easily hydrolyzed by the immobilized enzyme compared with synthetic substrates of low molecular weight. This may be due to attractive forces between the substrate and carrier.

On the other hand, in the case of enzymes originally acting on compounds of low molecular weight, a change in the substrate specificity of the enzyme on immobilization generally does not occur. For example, the authors found that mold aminoacylase immobilized by ionic binding with DEAE-cellulose or DEAE-Sephadex showed the same substrate specificity as the native enzyme toward various acyl-DL-amino acids.[13~15] L-Amino acid oxidase immobilized by diazo binding with porous glass[16] and D-amino acid oxidase immobilized by using CNBr-activated Sepharose[17] gave similar results.

In the case of immobilized microbial cells, the entrapping method using semipermeable polymer is generally adopted for immobilization, and the enzyme activity toward substrates of high molecular weight inevitably decreases.

3.2 pH DEPENDENCE OF ENZYME REACTIONS

As enzymes consist of protein, the catalytic activity is markedly affected by environmental conditions, especially the pH of aqueous media. Thus, information on changes in pH-activity behavior caused by immobilization of enzymes is useful for an understanding of the structure-function relationship of enzyme proteins.

3.2.1 Optimum pH of Immobilized Enzymes

Cases are known of change and no change in the optimum pH of an enzyme reaction and in the pH-activity curve upon immobilization.

A. Examples showing shifts in optimum pH

The changes in optimum pH and pH-activity curve depend on the charge of the enzyme protein and/or of the water-insoluble carrier. Examples of shifts of optimum pH on immobilization are listed in Table 3.2.

There are many cases showing shifts in optimum pH but not showing any change in pH-activity curve on immobilization, that is, a parallel shift of the pH-activity curve occurs without changing the pH-activity profile.

TABLE 3.2 Optimum pH's of immobilized enzymes

Enzyme	Immobilization method	Substrate	Optimum pH		Reference
			Native enzyme	Immobilized enzyme	
NADH dehydrogenase	cross-linking (glutaraldehyde)	NADH	6.0	5.5	18
Protocatechuate 3,4-dioxygenase	peptide binding (CNBr-activated Sepharose)	protocatechuic acid	8.4	7.9	19
Polynucleotide phosphorylase	diazo binding (porous glass)	ADP	9.0	9.4	20
Glucoamylase	alkylation (dichloro–s–triazinyl cellulose)	maltose	5.0	4.3	21
	cross-linking (glutaraldehyde)	maltose	5.0	4.3	22
β-Galactosidase	microcapsule (cellulose nitrate)	ONPG	7–7.5	6.5	23
Invertase	physical adsorption (activated carbon)	sucrose	4.5–5.0	5.5–6.0	24
	(bentonite)	sucrose	5.2	6.3	25
	ionic binding (DEAE-cellulose)	sucrose	5.4	3.4	26
	alkylation (monochloro–s-triazinyl DEAE-cellulose)	sucrose	4.5–5.0	4–4.5	24
	entrapping (polyacrylamide gel)	sucrose	4.5–5.0	5.5	24
		sucrose	4.5	4.0	27
Chymotrypsin	peptide binding (CM-cellulose azide)	casein	8.0	8.5	2
	(CNBr-activated Sephadex)	ATEE	8.0	9.5	28, 29
Trypsin	ionic binding (DEAE-Sephadex)	TAME	8.5	7.0	4
	peptide binding (CM-cellulose azide)	casein	8.5	9.0	2
	(ethylene-maleic anhydride copolymer)	BAEE	8.5	10.5	30

(*Continued*)

TABLE 3.2—*Continued*

Enzyme	Immobilization method	Substrate	Optimum pH Native enzyme	Optimum pH Immobilized enzyme	Reference
	(CNBr-activated Sepharose)	BAEE	8.4	9.0	31
		BAEE	8.4	9.5	31
		TAME	8.5	9.5	4
	(isothiocyanate-activated porous glass)	TAME	8.5	9.0	4
Papain	carrier cross-linking				
	(ZrO₂-porous glass +glutaraldehyde)	casein	6.5	7.0	32
Asparaginase	entrapping				
	(polyacrylamide gel)	L-Asn	8.0	7.0	33
Aminoacylase	ionic binding				
	(DEAE–cellulose)	Ac-DL-Met	7.5	7.0	13
	(DEAE–Sephadex)	Ac-DL-Met	7.5	7.0	15
	alkylation				
	(iodoacetyl cellulose)	Ac-DL-Met	7.5	8.0	34
	entrapping				
	(polyacrylamide gel)	Ac-DL-Met	7.5	7.0	35
Aminoacylase I	diazo binding				
	(Enzacryl AA)	Ac-DL-Met	7.0	7.5–8.0	36
ATP deaminase	ionic binding				
	(DEAE-cellulose)	ATP	5.0	3.0	37
Apyrase	peptide binding				
	(CM-cellulose azide)	ATP	6.4	7.4	38
	(CM-cellulose+ Woodward's reagent K)	ATP	6.4	7.4	38
	(polygalacturonic acid +Woodward's reagent K)	ATP	6.4	7.8	38
Aldolase	diazo binding				
	(PAB-cellulose)	FDP	8.0	6–6.5	39
	peptide binding				
	(ethylene-maleic anhydride copolymer)	FDP	8.0	8.5–9.0	39
	carrier cross-linking				
	(AE-cellulose+ glutaraldehyde)	FDP	8.0	6.5–7.0	39

ZrO_2, CO_2

For example, the optimum pH of aminoacylase immobilized by ionic binding with DEAE-cellulose or DEAE-Sephadex shifted 0.5 pH unit to the acid side compared to the native enzyme.[13,15] Similar shifts in optimum pH were observed with invertase[26] and ATP deaminase[37] immobilized by ionic binding with DEAE-cellulose, and NADH dehydrogenase[18] immobilized by cross-linking using glutaraldehyde.

These shifts toward the acid side may be explained as follows. When an enzyme is bound to a polycationic carrier, positive charge on the enzyme protein increases, and the pH of the immobilized enzyme region becomes more alkaline than that of the external solution. Accordingly, the enzyme reaction effectively proceeds on the alkaline side of the external buffer pH, and the optimum pH apparently shifts to the acid side.

On the other hand, the optimum pH's of trypsin,[2] chymotrypsin,[2] and apyrase[38] immobilized by peptide binding with CM-cellulose azide shifted 0.5~1 pH unit to the alkaline side compared to the native enzyme. This shift to the alkaline side may be caused by an increase of negative charge due to carboxyl groups of CM-cellulose. That is, the microenvironment of the immobilized enzyme becomes more acidic than that of the external solution.

Goldstein et al.[40,41] carried out experiments to investigate this pH shift in relation to the electrostatic field produced by a highly charged carrier. As shown in Fig. 3.2, the optimum pH of chymotrypsin immobi-

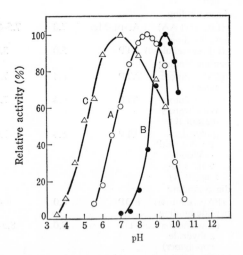

Fig. 3.2   pH-Activity curves at low ionic strength for chymotrypsin.[40] Substrate, acetyl-L-tyrosine ethyl ester; ionic strength ($\Gamma/2$), 0.008.
A, chymotrypsin; B, ethylene-maleic anhydride copolymer-chymotrypsin; C, polyornithyl-chymotrypsin.

lized by using a polyanionic carrier, ethylene-maleic anhydride copolymer, shifted 1 pH unit to the alkaline side, while in chymotrypsin immobilized by using a polycationic carrier, polyornithine, a shift of 1.5 pH unit to the acid side occurred. Further, the pH-activity curve for immobilized trypsin (IMET) prepared by using ethylene-maleic anhydride copolymer at various ionic strengths was studied, and it was found that the curve became similar to that of the native enzyme with increasing ionic strength (Fig. 3.3). That is, ethylene-maleic anhydride copolymer has negatively charged

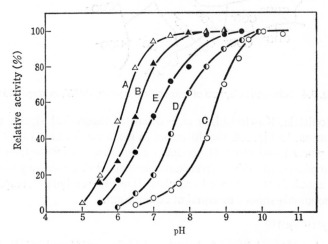

Fig. 3.3   pH-Activity curves at various ionic strengths for trypsin.[41)]
Trypsin was immobilized using ethylene-maleic anhydride copolymer.
Substrate, benzoyl-L-arginine ethyl ester.
Ionic strength ($\Gamma/2$) for native enzyme: A, 3.5×10^{-2}; B, 1.0. Ionic strength ($\Gamma/2$) for immobilized enzyme: C, 6.0×10^{-3}; D, 3.5×10^{-2}; E, 1.0.

carboxyl groups, which surround trypsin as shown in Fig. 3.4[41)] At equilibrium, an electro-statically neutral state should be maintained. Thus, in order to counteract negative charges of carboxyl groups, the concentration of cations, i.e., H^+ and Na^+, in the internal region of the carrier should be higher than that in the external region. In other words, the pH of the internal region of the carrier should be low. As pH-activity curves are expressed based on the pH of the external solution, even if the optimum pH of IMET is same as that of the native enzyme, the optimum pH apparently shifts to the alkaline side because the pH of the external solution is higher than that of the internal region. At higher ionic strength, as negative charges of the carrier are neutralized by Na^+ rather than H^+, the difference of pH between external and internal regions becomes small, and the shift of optimum pH appears to be small.

phase I   phase II

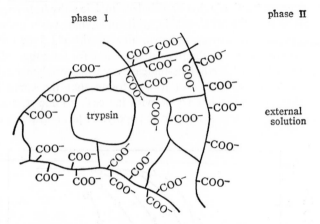

external
solution

Fig. 3.4   Schematic diagram of immobilized trypsin, IMET, in suspension.[41]

Goldstein, Katchalski, *et al.*[41] discussed theoretical aspects of these phenomena. In Fig. 3.4, the difference in mean electrostatic potentials between phase I and phase II produces a difference in ion distributions in the two phases, and this difference is expressed as Ψ. Expressing the chemical potentials of H^+ in phase I and phase II as μH^{+I} and μH^{+II}, respectively, both potentials should be equal at equilibrium.

$$\mu H^{+I} = \mu H^{+II} \tag{3.1}$$

Employing H^+ activities of phase I and II, i.e., aH^{+I} and aH^{+II}, Eqs. 3.2 and 3.3 can be derived.

$$\mu H^{+I} = \mu H^{+0} + kT \ln aH^{+I} + e\Psi \tag{3.2}$$

$$\mu H^{+II} = \mu H^{+0} + kT \ln aH^{+II} \tag{3.3}$$

In Eq. 3.2, e is the positive electron charge. These equations can be rewritten as Eq. 3.4.

$$\left. \begin{array}{l} \ln aH^{+II} - \ln aH^{+I} = e\Psi/kT \\ \Delta pH = pH^{I} - pH^{II} = 0.43\, e\Psi/kT \end{array} \right\} \tag{3.4}$$

From Fig. 3.3, the values of ΔpH, i.e., differences of pH values at which the native and immobilized enzymes show half-maximal activities at a specified ionic strength, are determined, and the values of Ψ calculated from Eq. 3.4 are shown in Table 3.3. The value of Ψ at $\Gamma/2 = 6 \times 10^{-3}$ in Table 3.3 is -145 mV, and this is in fairly good agreement with the value of Ψ for ethylene-maleic anhydride copolymer, -160 mV, at the same ionic strength, indicating that this theory is applicable to immobilized enzyme systems.

TABLE 3.3   Effect of ionic strength on the pH-activity profile
of immobilized trypsin[41]

Ionic strength ($\Gamma/2$)	$-\Delta$pH	$-\Psi$† ($V \times 10^3$)
0.6×10^{-2}	2.4 ± 0.2	145 ± 15
1.0×10^{-2}	2.0	124 ± 13
3.5×10^{-2}	1.6	96 ± 12
0.2	1.3	81 ± 10
1.0	0.4	27 ± 8

† Ψ was calculated from Fig. 3.3 using Eq. 3.4.

Further, the optimum pH of chymotrypsin[29] immobilized by using CNBr-activated Sephadex shifted 2 pH units toward the alkaline side compared to the native enzyme. When the carrier, Sephadex, was decomposed by means of dextranase, the optimum pH of the resulting preparation approached that of the native enzyme. This shift of optimum pH can be explained by accumulation of H^+ in internal regions of the Sephadex gel.

There have also been some reports that both a shift of optimum pH and a change of the pH-activity curve occur. In glucoamylase[21] immobilized by alkylation with dichloro-*s*-triazinyl cellulose, the optimum pH shifted to the acid side, and the pH-activity curve became narrow compared with that of the native enzyme.

In the case of immobilized enzymes prepared by entrapping in

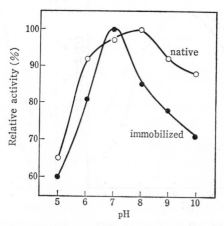

Fig. 3.5   Effect of pH on the reaction rates of native and immobilized asparaginase.[33] Asparaginase was immobilized in polyacrylamide gel. Substrate, L-asparagine.

polyacrylamide gel, modification of the enzymes should not occur theo-
retically, and polyacrylamide gel is electrically neutral. Nevertheless, a
shift of optimum pH and change of pH-activity curve are occasionally
observed in this type of immobilized enzyme. For example, in the cases
of immobilized asparaginase[33] (Fig. 3.5), aminoacylase,[35] and phospho-
glyceromutase,[42] optimum pH values shifted to the acid side and the pH-
activity curves became narrower.

Further, there is an interesting report that the optimum pH of immobilized trypsin prepared in the presence of substrate differed from that
prepared in the absence of substrate.[31]

B.   Examples showing no shift in optimum pH

Some immobilized enzymes show no shift of optimum pH but do
exhibit changes of pH-activity curve.

For example, in β-fructofuranosidase[43] (Fig. 3.6) immobilized by
diazo binding with polyaminopolystyrene, and asparaginase[44] microencapsulated in nylon or polyurea, the optimum pH did not shift, but
the pH-activity curves became narrower. The pH-activity curve of enolase
entrapped in polyacrylamide gel became broader, as shown in Fig. 3.7.[45]

On the other hand, both the optimum pH and pH-activity curve of
immobilized enzymes are the same as those of the native enzymes in the
following cases: immobilized chymotrypsin,[46] trypsin,[47] and papain[8]

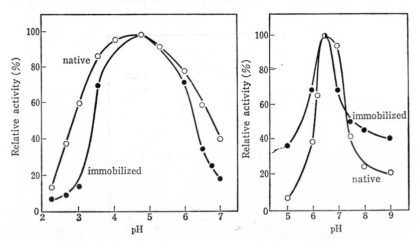

Fig. 3.6 Effect of pH on the reaction rates
of native and immobilized β-fructofuranosid-
ase.[43] Substrate, sucrose.

Fig. 3.7 Effect of pH on the
reaction rates of native and
immobilized enolase.[45] Substrate, 2-phosphoglycerate.

prepared by diazo binding with amino acid copolymer; immobilized ribonuclease[48] and thrombokinase[49] prepared with CNBr-activated Sepharose; and immobilized aspartase[50] prepared by entrapping in polyacrylamide gel.

3.2.2 Optimum pH of Immobilized Microbial Cells

The optimum pH of immobilized microbial cells has also been studied and compared with that of intact cells or native enzyme.

In *Escherichia coli* and *Azotobacter agile* immobilized by using Dowex 1 (Cl type), the optimum pH for oxidation of glucose or succinic acid shifted 1 pH unit to the alkaline side.[51,52]

The authors also reported that the optimum pH values of aspartase activity of *Escherichia coli*[53] and fumarase activity of *Brevibacterium ammoniagenes*[54] immobilized by entrapping in polyacrylamide gel shifted to the acid side compared with those of intact cells. No shift of optimum pH was observed with L-histidine ammonia-lyase activity of *Achromobacter liquidum*,[55] L-arginine deiminase activity of *Pseudomonas putida*,[56] and penicillin amidase activity of *Escherichia coli*[57] immobilized by the same entrapping method.

As mentioned above, in certain cases, some regularities are observed in the changes of optimum pH and pH-activity curve caused by the immobilization of enzymes or microbial cells. Accordingly, when the optimum pH of a native enzyme is different from the pH desired for a practical enzyme reaction, it may be possible to shift the optimum pH to the desired value for practical use by selecting an appropriate immobilization method.

3.3 TEMPERATURE DEPENDENCE OF ENZYME REACTIONS

The catalytic activity of enzymes is dependent on temperature, as in the case of ordinary chemical catalysts, but the activity is lost at temperatures above a certain limit due to the denaturation of enzyme protein. Therefore, it is considered that studies on temperature dependence caused by immobilization may provide useful information for comparisons of enzymes and chemical catalysts.

3.3.1 Optimum Temperatures of Immobilized Enzymes

Upon the immobilization of enzymes, changes in the optimum temperature of the enzyme reaction occur in some cases.

The optimum temperatures of immobilized enzymes were 5–15°C higher than those of native enzymes in the cases of trypsin[2,58] and chymotrypsin[2] immobilized by peptide binding with CM-cellulose azide, and tryptophanase[59] immobilized by covalent binding with pyridoxal phosphate-Sepharose derivative. Elevation of optimum temperature was also observed in enolase[60] and aminoacylase[35] immobilized by entrapping in polyacrylamide gel.

As shown in Fig. 3.8, the authors reported that the optimum temperatures of aminoacylases immobilized by ionic binding with DEAE-cellulose or DEAE-Sephadex, and by entrapping in polyacrylamide gel became higher than that of native enzyme, while in aminoacylase immobilized by alkylation with iodoacetyl-cellulose, the optimum temperature fell. In lysozyme[61] immobilized by physical adsorption on collagen membrane, asparaginase[44] microencapsulated in nylon or polymer membrane, and NADH dehydrogenase[18] immobilized by cross-linking with glutaraldehyde, decreases of optimum temperature were observed. In the case of invertase, immobilized enzymes prepared by various methods all showed lower optimum temperatures, as shown in Table 3.4.

On the other hand, there have been many reports that no change of optimum temperature occurs on immobilization. For example, in the cases of glucose isomerase[62] and leucine aminopeptidase[63] immobilized by diazo binding with porous glass, chymotrypsin[64] immobilized with

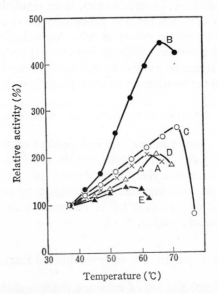

Fig. 3.8   Effect of temperature on the reaction rate of aminoacylase.[15,34,35] Substrate, acetyl-DL-methionine; reaction time, 30 min. A, native aminoacylase; B, DEAE-cellulose-aminoacylase; C, DEAE-Sephadex-aminoacylase; E, iodoacetylcellulose-aminoacylase.

TABLE 3.4   Optimum temperatures of various immobilized
invertases[24]

	Enzyme preparation	Optimum temperature (°C)
Native invertase		55
Immobilized invertase		
①	Physical adsorption	
	Activated carbon	40
②	Ionic binding	
	DEAE-cellulose	30
③	Alkylation	
	Monochloro–*s*-triazinyl DEAE-cellulose	45
④	Entrapping	
	Polyacrylamide gel	50

CNBr-activated cellulosic membrane, trypsin[65] immobilized with CNBr-activated agarose, and asparaginase[33] immobilized by entrapping in polyacrylamide gel, the optimum temperatures were same as those of the native enzymes.

3.3.2   Activation Energy of Immobilized Enzymes

Various studies on the activation energy of immobilized enzymes have been reported.

The activation energies of immobilized enzymes were almost the same as those of the corresponding native enzymes in the case of invertase[24] immobilized by physical adsorption on activated carbon, asparaginase[66] and invertase[24,67] immobilized by ionic binding with DEAE-cellulose, asparaginase[44] microencapsulated in nylon or polyurea membrane, and invertase[24] immobilized by entrapping in polyacrylamide gel. Similar results were obtained with lactase,[68] invertase[24] and glucoamylase[21] immobilized by alkylation with chloro-*s*-triazinylcellulose or chloro-*s*-triazinyl DEAE-cellulose, and amylase[1] immobilized by peptide binding with CM-cellulose azide.

On the other hand, the activation energy of lysozyme[61] immobilized by physical adsorption on collagen membrane increased compared to the native enzyme. An opposite result, decrease of activation energy, was observed with aminoacylase[36] immobilized by diazo binding with Enzacryl AA.

The authors prepared immobilized aminoacylases by various methods, and determined the activation energies. As shown in Table 3.5, the activation energies of aminoacylases[13,15] immobilized by ionic binding with

TABLE 3.5　Activation energies of various immobilized aminoacylases[13,15,34,35]

Enzyme preparation	Activation energy (cal/mol)
Native aminoacylase	6,700
Immobilized aminoacylase	
① Ionic binding	
DEAE–cellulose	11,100
DEAE–Sephadex	7,640
② Alkylation	
Iodoacetylcellulose	3,900
③ Entrapping	
Polyacrylamide gel	5,700

DEAE-cellulose and DEAE-Sephadex increased compared to the native enzyme, while decreases of activation energy were observed with aminoacylases immobilized by alkylation with iodoacetylcellulose[34] and by entrapping in polyacrylamide gel.[35]

3.3.3　Optimum Temperature and Activation Energy of Immobilized Microbial Cells

Optimum temperatures and activation energies of immobilized microbial cells have been also measured, and compared with those of intact cells.

The authors found that the optimum temperatures of immobilized cells were the same as those of intact cells in the cases of aspartase activity of *Escherichia coli*[53], fumarase activity of *Brevibacterium ammoniagenes*,[54] penicillin amidase activity of *Escherichia coli*,[57] and L-histidine ammonia-lyase activity of *Achromobacter liquidum*[55] immobilized by entrapping in polyacrylamide gel. However, an increase in optimum temperature of 20°C was observed in L-arginine deiminase activity of *Pseudomonas putida*[56] immobilized by the above method.

Further, the temperature dependence of lipase activity of *Pseudomonas mephitica* var. *lipolytica* immobilized by heat and pH treatment was studied, and the activation energies were calculated to be 21,000 cal/mol and 15,000 cal/mol for immobilized and intact cells, respectively.[69]

3.4　KINETIC CONSTANTS

As conformational changes of the enzyme protein may occur on immobilization, and the affinity between enzyme and substrate may be changed, the investigation of kinetic constants for immobilized enzymes is very important.

3.4.1 Michaelis Constant

As shown in Table 3.6, little or no change occurred in Michaelis constant (K_m), which reflects the affinity between enzyme and substrate, on immobilization in some cases, though significant changes of up to two orders of magnitude occur in a few cases.

In the case of the carrier binding method, electrostatic interaction between the carrier and substrate is considered to be one of the reasons for changes of K_m value upon immobilization.

The K_m value for ficin[10] immobilized by peptide binding with CM-cellulose azide decreased by 1/10 compared with that of the native enzyme, when benzoyl-L-arginine ethyl ester (BAEE) was used as a substrate. This decrease can be explained in terms of an increase of electrostatic attractive force between the polyanionic carrier, CM-cellulose, and positively charged substrate, BAEE, so that the substrate concentration in the immobilized enzyme region becomes higher than that in the external solution. Thus, the enzyme reaction effectively proceeds at a higher substrate concentration than the substrate concentration used, and K_m apparently decreases. Conversely, K_m for creatine kinase[72] immobilized by peptide binding with CM-cellulose azide became ten times higher than that of the native enzyme due to repulsion between the polyanionic carrier and negatively charged ATP or creatine.

The K_m value for trypsin immobilized by using a polyanionic carrier, ethylene-maleic anhydride copolymer, was measured by using positively charged benzoyl-L-arginine amide (BAA) as a substrate at various ionic strengths.[41] As shown in Fig. 3.9, K_m for the immobilized enzyme decreased to 1/10 of that of the native enzyme at lower ionic strength, but it approached that of native enzyme at higher ionic strengths. Thus, the change in K_m value is considered to be caused by electrostatic interaction between the carrier and substrate.

In order to explain the change of K_m value upon immobilization of enzymes, theoretical analysis was carried out as in the case of optimum pH shift. That is, by employing substrate concentrations, C_s^I and C_s^{II}, in place of H$^+$ activity in Eq. 3.4, Eq. 3.5 is obtained, and can be rewritten as Eq. 3.6. When a substrate is positive and a carrier is negative, the

$$\ln Cs^{II} - \ln Cs^I = Ze\Psi/kT \qquad (3.5)$$

$$Cs^I = Cs^{II} \exp(-Ze\Psi/kT) \qquad (3.6)$$

number of charges, Z, is positive and Ψ is negative, so that $C_s^I > C_s^{II}$. By substituting Eq. 3.6 into the Michaelis-Menten equation, Eq. 3.7 is obtained; apparent K_m is given by $K_m \exp(Ze\Psi/kT)$.

TABLE 3.6 Michaelis constants of immobilized enzymes

Enzyme	Immobilization method	Substrate	K_m (mM) Native enzyme	K_m (mM) Immobilized enzyme	Reference
Lactate dehydrogenase	diazo binding (porous glass)	lactate	8.0	5.9	70
	carrier cross-linking	pyruvate	0.125	0.39	70
	(porous glass)	lactate	8.0	8.0	70
		pyruvate	0.125	0.4	70
Glucose oxidase	diazo binding (porous glass)	D-glucose	7.7	6.8	71
Glyceraldehyde-phosphate dehydrogenase	carrier cross-linking (AE-cellulose+ glutaraldehyde)	GAP	6.4	0.66	39
L-Amino acid oxidase	diazo binding (porous glass)	L-Leu	1.0	4.0	71
NADH dehydrogenase	cross-linking (glutaraldehyde)	NADH	6.7×10^{-2}	31×10^{-2}	18
Creatine kinase	peptide binding (CM-cellulose azide)	ATP	0.65	7.0	72
Alkaline phosphatase	diazo binding (porous glass)	PNPP	0.1	2.9	71
Arylsulfatase	diazo binding (porous glass)	PNPS	1.85	1.57	71
Glucoamylase	peptide binding (CM-cellulose azide)	amylose	1.4×10^{-4}	1.1×10^{-4}	1
	alkylation (bromoacetyl-cellulose)	maltose	0.9	1.35	73
β-Galactosidase	alkylation (dichloro–s–triazinyl–cellulose)	ONPG	0.2	0.8	68
Aminopeptidase M	diazo binding (porous glass)	LpNA	0.28	1.07	63
Chymotrypsin	peptide binding (CM-cellulose azide)	ATA	42	25	2
	(CNBr-activated Sepharose)	ATEE	3.3	30	29
Trypsin	ionic binding (DEAE–Sephadex)	TAME	0.12	2.5	4
	peptide binding CNBr-activated Sepharose	TAME	0.12	6.2	4
		BAEE	0.4×10^{-2}	0.41	31
		BAEE	0.4×10^{-2}	0.24	31
	(isothiocyanate-activated porous glass)	TAME	0.12	1.53	4

(*Continued*)

TABLE 3.6—*Continued*

Enzyme	Immobilization method	Substrate	K_m (mM) Native enzyme	K_m (mM) Immobilized enzyme	Reference
	(ethylene-maleic anhydride copolymer)	BAA	6.85	0.2	41
	carrier cross-linking (porous glass + glutaraldehyde)	BAEE	1.2×10^{-2}	0.285	74
Ficin	peptide binding (CM-cellulose azide)	BAEE	20	2.0	10, 72
Asparaginase	entrapping (polyacrylamide gel)	L-Asn	1.8×10^{-2}	3.3	33
	(nylon micro-capsules)	L-Asn	1.8×10^{-2}	1.3	44
	(polyurea micro-capsules)	L-Asn	1.8×10^{-2}	2.0	44
Urease	entrapping (liquid membrane microcapsules)	urea	3.4	1.8×10^2	75
Aminoacylase	ionic binding (DEAE-cellulose)	Ac–DL-Met	5.7	3.5	13
	(DEAE–Sephadex)	Ac–DL-Met	5.7	8.7	15
	alkylation (iodoacetyl cellulose)	Ac–DL-Met	5.7	6.7	34
	entrapping (polyacrylamide gel)	Ac–DL-Met	5.7	5.0	35
Apyrase	peptide binding (CM-cellulose azide)	ATP	9×10^{-2}	0.21	76
	(polyaspartic acid + Woodward's reagent K)	ATP	9×10^{-2}	0.67	38
Aldolase	diazo binding (PAB-cellulose)	FDP	4×10^{-2}	6×10^{-3}	39
	peptide binding (ethylene-maleic anhydride copolymer)	FDP	4×10^{-2}	3.6	39
Citrate synthase	peptide binding (CNBr-activated Sephadex)	OAA	2.1×10^{-3}	8.7×10^{-3}	77
	entrapping (polyacrylamide gel)	OAA	2.1×10^{-3}	3.2×10^{-2}	77
Carbonic anhydrase	entrapping (nylon microcapsules)	PNPA	13	36	78

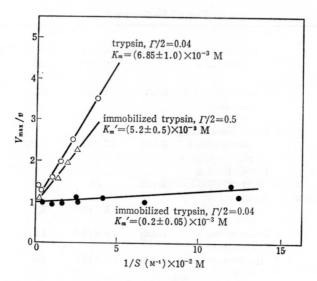

Fig. 3.9   Lineweaver-Burk plots for trypsin.[41]
Trypsin was immobilized using ethylene-maleic anhydride copolymer.
Substrate, benzoyl-L-arginine amide.

$$-Vs = \frac{k_2 E_0 C_S^{II}}{K_m \cdot \exp(Ze\varPsi/kT) + C_s^{II}} \qquad (3.7)$$

Accordingly, when the charges of the carrier and substrate are opposite, i.e., $Z\varPsi < 0$, apparent K_m decreases, while when the charges are
same, i.e., $Z\varPsi > 0$, apparent K_m increases. When either one of the charges
is 0, Z or $\varPsi$ becomes 0, and the K_m value does not change. As the ionic
strength increases, the absolute value of $\varPsi$ becomes small, and the effect
of charge decreases.

As well as the carrier charge described above, changes in K_m value
caused by other factors have been reported.

For instance, in the case of chymotrypsin immobilized by CNBr-
activated Sephadex, K_m for acetyl-L-tyrosine ethyl ester (ATEE) was
calculated to be 25–30 mM. This value was ten times higher than that of
the native enzyme.[29] When this immobilized enzyme was treated with
dextranase to partially hydrolyze Sephadex, K_m for the treated enzyme
preparation was 5 mM, approaching that of the native enzyme. In this
case, the change in K_m upon immobilization was considered to be caused
not by electrostatic interaction between the carrier and substrate, but by
diffusion of the substrate and product in the carrier, which became limiting and resulted in an apparent increase of K_m.

Immobilized glucoamylases of various particle sizes were prepared by alkylation with bromoacetylcellulose, and the relation between K_m value and particle size was studied.[73] K_m values for maltose were calculated to be 0.9×10^{-3} M in the case of the native enzyme, and 1.35×10^{-3} M, 1.6×10^{-3} M and 2.15×10^{-3} M in the cases of the immobilized enzymes with particle sizes of below 15 μ, 15–55 μ and 70–190 μ, respectively. That is, the apparent K_m of the immobilized enzyme decreased with smaller particle size. In this case, too, the effect of internal diffusion of the substrate and product on the reaction rate became less and K_m approached that of the native enzyme.

The authors measured K_m values for asparaginases microencapsulated in nylon or polyurea membrane,[44] and immobilized by entrapping in polyacrylamide gel.[33] K_m values for these immobilized asparaginases were found to become higher by two orders of magnitude than that of the native enzyme. This increase of K_m may be explained in terms of a decrease of substrate concentration in the gel. That is, permeation of substrate through the microcapsule membrane or gel lattice became limiting, and the internal substrate concentration became lower than that in the outer region, resulting in an increase of apparent K_m.

3.4.2 Maximum Reaction Velocity

The maximum reaction velocity (V_{max}) of immobilized enzymes has also been measured.

As shown in Table 3.7, the authors determined V_{max} values for aminoacylases immobilized by ionic binding with DEAE-Sephadex, by alkylation with iodoacetylcellulose, and by entrapping in polyacrylamide gel. The values obtained were almost the same as that for the native enzyme. Similar results were obtained with invertase[79] immobilized by diazo binding with porous glass and subtilisin Novo[80] immobilized with CNBr-activated Sepharose.

TABLE 3.7 Maximum reaction velocity of various immobilized aminoacylases[15,34,35]

Enzyme preparation	Maximum reaction velocity (μmol/h)
Native aminoacylase	1.52
Immobilized aminoacylase	
① Ionic binding	
DEAE–Sephadex	3.33
② Alkylation	
Iodoacetylcellulose	4.65
③ Entrapping	
Polyacrylamide gel	2.33

Further, V_{max} for β-galactosidase[68] immobilized by alkylation with dichloro-s-triazinylcellulose was about 30 times higher than that of the native enzyme, while in the cases of β-fructofuranosidase[81] immobilized by diazo binding with polyamino styrene, and invertase[82] immobilized by entrapping in polyacrylamide gel, the V_{max} values decreased to 1/10 of those of native enzymes.

3.5 STABILITY

As a result of the immobilization of enzymes, increases of stability have been observed in many cases. The enhancement of stability is advantageous for the industrial application of immobilized enzymes, and it is thus important in determining the feasibility of immobilized enzymes for a particular application. Therefore, the stabilities of various kinds of immobilized enzymes have been widely investigated.

Merlose[83] investigated 50 reported immobilized enzymes and compared their stabilities with those of the native enzymes. Of the 50 instances, stability was increased by immobilization in 30 cases, decreased in only 8 cases, and there was almost no change in 12 cases. However, there is a tendency for stabilized cases to be more generally reported than cases showing decreased stability. Accordingly, Merlose's results are not necessarily applicable in general, but nevertheless it can be expected that an enzyme may be stabilized by appropriate immobilization. However, as the stabilities of immobilized enzymes have been determined under various conditions, and it is difficult to carry out accurate comparisons, no clear correlation between stability and immobilization method has yet been found.

3.5.1 Stability toward Reagents

Immobilization of enzymes also induces changes in the stability of enzymes toward various reagents. Some examples are given in Table 3.8.

Enhancement of stability toward protein-denaturing agents or enzyme inhibitors has been observed in some immobilized enzymes. In particular, immobilized aminoacylase[15] and immobilized trypsin[2,88] were not inhibited by urea, but activated. Immobilized aminoacylase was also activated by guanidine or n-propanol. The activation mechanism is not clear, but may be due to conformational change of the protein, e.g., conversion of the rigid structure of the immobilized aminoacylase into a more flexible structure suitable for catalytic action.

Enhanced resistance toward inhibitors was also reported with some immobilized enzymes. Immobilized trypsins prepared by peptide bind-

TABLE 3.8   Stability of immobilized enzymes toward various reagents

Enzyme	Immobilization method	Reagent	Remaining activity		Reference
			Native enzyme	Immobilized enzyme	
Lactate dehydrogenase	diazo binding (porous glass)	antibody	0	20	70
	carrier cross-linking (porous glass+ glutaraldehyde)	antibody	0	10	70
Protocatechuate 3,4-dioxygenase	peptide binding (CNBr–activated Sepharose)	6 M urea	0	20	19
Acetylcholin-esterase	entrapping (silastic resin)	bidrin[†]	0	49	84
α-Amylase	peptide binding (CNBr–activated Sephadex)	6 M urea	25	45	85
	(CNBr–activated Sepharose)	1 mM EDTA	5	30	85
		6 M urea	25	74	85
Glucoamylase	physical adsorption (activated carbon)	0.002 M mercuric acetate	80	0	86
	alkylation (bromoacetyl-cellulose)	6 M urea	25	0 (70–190 μ)	73
				25 (<15 μ)	73
Leucine aminopeptidase	physical adsorption (calcium phosphate gel)	30% acetone	100	80	87
		30% ethanol	90	70	87
		30% dioxane	60	0	87
Chymotrypsin	peptide binding (CNBr–activated cellulosic membrane)	4 M urea	50	80	64
		7 M urea	0	35	64
Trypsin	peptide binding (CM-cellulose azide)	3 M urea	60	120	2
		soybean trypsin inhibitor	0	80	2
	(CNBr–activated Sepharose)	8 M urea	0	60	88
	carrier cross-linking (AE-cellulose+ glutaraldehyde)	soybean trypsin inhibitor	0	50–60	89
Aminoacylase	ionic binding (DEAE–Sepha-dex)	6 M urea	9	146	15
		2 M guanidine	49	117	15
		1% SDS	0	35	15
		24% n-propanol	55	138	15

†
$$(CH_3O)_2\text{-}P\text{-}O\text{-}C = C\text{-}C\text{-}N\text{-}(CH_3)_2$$
with O (‖) on P, and CH₃, H, O substituents: $(CH_3O)_2\overset{O}{\overset{\|}{P}}\text{-O-}\overset{CH_3}{\overset{|}{C}}=\overset{H}{\overset{|}{C}}\text{-}\overset{O}{\overset{\|}{C}}\text{-N-}(CH_3)_2$

ing[2] and carrier cross-linking[89] with CM-cellulose azide were found to become resistant to trypsin inhibitor. It may be considered that the approach of high-molecular-weight trypsin inhibitor to the immobilized enzyme is interfered with by steric hindrance, as in the cases of immobilized proteolytic enzymes and a high-molecular-weight substrate, casein.

It was reported that myosin ATPase was activated by Ca^{2+} and inhibited by Mg^{2+}, while the immobilized enzyme prepared by diazo binding with PAB-cellulose was activated by both ions.[90] In lactate dehydrogenase immobilized by glutaraldehyde on porous glass, enhanced resistance toward substrate inhibition was observed.[91]

On the other hand, there have been some examples of decreases of stability upon immobilization. For instance, glucoamylase immobilized by physical adsorption on activated carbon showed enhanced inhibition by mercury acetate. There has been an interesting study showing that in the case of glucoamylase immobilized by alkylation with bromoacetylcellulose, the stability toward urea was dependent on the particle size of the immobilized enzyme; it decreased with increase of the particle size.[73]

Further, some cases showing no change of stability toward protein-denaturing agents or enzyme inhibitors upon immobilization have been reported. In the case of amylase immobilized by peptide binding with CM-cellulose azide, the stability toward protein-denaturing agents did not change.[1] The effects of catalase inhibitors such as cyanide, hydroxylamine and aminotriazole on immobilized catalase prepared by cross-linking with glutaraldehyde were the same as on the native enzyme.[92] In the case of trypsin immobilized by glutaraldehyde on AE-cellulose, the enzyme became less sensitive to inhibition by soybean trypsin inhibitor of high molecular weight compared with the native enzyme, while inhibition by a low-molecular-weight inhibitor, arginine, did not change.[89]

As mentioned above, the stabilities toward various reagents are important in considering the applications of immobilized enzymes. Such stabilization of enzymes upon immobilization will clearly be the subject of much study. For instance, it has been reported that some immobilized enzymes show enhanced stability toward various organic solvents in which the native enzymes are unstable.[15] This raises the possibility that enzyme reactions can be carried out in organic solvents, which would greatly increase the utility of immobilized enzymes as catalysts for organic synthetic chemistry.

3.5.2 Stability toward Proteolytic Enzymes

In enzyme reactions using a crude enzyme preparations, inactivation of the enzyme is often accelerated by contaminating proteolytic enzymes.

However, as shown in Table 3.9, in some cases resistance to various proteolytic enzymes is increased by immobilization, and this is very advantageous.

The authors have shown that in the case of aminoacylase immobilized by ionic binding to DEAE-cellulose or DEAE-Sephadex, resistance to proteolytic enzymes such as trypsin increased compared with the native enzyme.[15] Similar phenomena were observed in tRNA nucleotidyltransferase[94] or isocitrate dehydrogenase[93] immobilized with CNBr-activated Sepharose. This stabilization was considered to be caused by steric hindrance of the approach of proteolytic enzymes of high molecular weight to the immobilized enzymes.

In the case of proteolytic enzymes such as trypsin and papain, the stability was enhanced by a decrease of catalytic activity toward substrates

TABLE 3.9   Stability of immobilized enzymes toward proteolytic enzymes

Enzyme	Immobilization method	Proteolytic enzyme	Remaining activity		Reference
			Native enzyme	Immobilized enzyme	
Lactate dehydrogenase	diazo binding (porous glass)	subtilisin	55	80	70
	carrier cross-linking (porous glass + glutaraldehyde)	subtilisin	55	100	70
Isocitrate dehydrogenase	peptide binding (CNBr–activated Sepharose)	trypsin	0	90	93
tRNA nucleo-tidyltransferase	peptide binding (CNBr–activated Sepharose)	trypsin	3	55	94
Asparaginase	entrapping (nylon micro-capsules)	trypsin	28	104	44
		chymotrypsin	4	111	44
		Pronase-P	12	106	44
	(polyurea micro-capsules)	trypsin	28	100	44
		chymotrypsin	4	96	44
		Pronase-P	2	107	44
	(polyacrylamide gel)	trypsin	14	109	33
		chymotrypsin	18	102	33
		Pronase-P	0	117	33
Aminoacylase	ionic binding (DEAE-cellulose)	trypsin	23	33	15
		Pronase-P	68	53	15
	(DEAE–Sepha-dex)	trypsin	23	87	15
		Pronase-P	68	88	15

of high molecular weight (i.e., the enzyme itself) due to steric hindrance, that is, by reducing the extent of autolysis.[8,80,95,96]

The authors also showed that immobilized asparaginases prepared by the entrapping method in nylon or polyurea membrane, and in polyacrylamide gel were very stable to proteolytic enzymes under conditions such that the native enzyme was almost completely inactivated.[33,34,97] In the case of enzymes immobilized by entrapping, the enzyme is entrapped in a semipermeable polymer, and does not come into contact with external proteolytic enzymes of high molecular weight. Accordingly, the entrapping method is considered to be effective for stabilizing enzymes with respect to proteolytic activity.

3.5.3 Heat Stability

The catalytic activity of enzymes increases with elevation of temperature, as in the case of usual chemical catalysts. Howeve, as enzymes consist of protein and are generally unstable to heat, the enzyme reaction cannot be practically carried out at high temperature. If the heat stability of an enzyme is enhanced by immobilization, the potential utilization of such enzymes will be extensive.

The heat stability of many immobilized enzymes has been studied, and there are examples showing increases, no change, and decreases of heat stability on immobilization. As shown in Table 3.10, enhancement of heat stability on immobilization has been observed with many enzymes.

The authors found that the heat stabilities of mold aminoacylases immobilized by ionic binding with DEAE-Sephadex, alkylation with iodoacetylcellulose, and entrapping with polyacrylamide gel increased considerably compared with that of the native enzyme, as shown in Fig. 3.10. For example, the native aminoacylase was almost completely inactivated by heat treatment at 70°C for 10 min, while the remaining activities of DEAE-Sephadex-aminoacylase, iodoacetylcellulose-aminoacylase, and polyacrylamide gel-aminoacylase after similar treatments were 90%, 50%, and 30%, respectively. Enhancement of heat stability was also observed in pronases immobilized by using ethylene-maleic anhydride copolymer and CNBr-activated cellulose,[102] and in acid phosphatase immobilized by carrier cross-linking with glutaraldehyde and polyacrylamide gel.[104] Further, an interesting study suggested that the heat stability of immobilized enzyme depended on the pore size of the carrier used for immobilization.[85] In this case, a similar tendency was also observed in other stabilities, i.e., the stabilities to pH, EDTA, and urea. Thus, it was suggested that there might be an optimum matrix size to stabilize the enzyme.

On the other hand, some examples showing decreases of heat stability

TABLE 3.10 Heat stabilities of immobilized enzymes

Immobilization method	Enzyme	Treatment		Remaining activity		Reference
		Temp. (°C)	Time (min)	Native enzyme	immobilized enzyme	
Physical adsorption						
Calcium phosphate gel	leucine aminopeptidase	55	10	100	30	87
Ionic binding						
DEAE-cellulose	aminoacylase	70	10	5	80	15, 98
DEAE-Sephadex	aminoacylase	70	10	5	90	15, 98
Diazo binding						
Enzacryl AA	aminoacylase I	60	10	20	80	36
Porous glass	lactate dehydrogenase	75	30	40	80	70
Peptide binding						
CNBr-activated	glucose 6-phosphate dehydrogenase	50	15	0	60	99
Sepharose	*t*RNA nucleotidyl-transferase	55	15	5	50	94
	propanediol dehydratase	60	5	5	50	100
CNBr-activated cellulose	dextranase	50	60	20	80	101
	pronase	55	120	0	50	102
Acrylamide-methacrylic acid copolymer azide	trypsin	100	10	87	23	103
Ethylene-maleic anhydride copolymer	pronase	55	120	0	33	102
Alkylation						
Iodoacetylcellulose	aminoacylase	70	10	5	50	98
Carrier cross-linking						
Polyacrylamide+glutaraldehyde	acid phosphatase	60	10	25	47	104
Porous glass+glutaraldehyde	lactate dehydrogenase	75	30	40	90	70
Entrapping						
Polyacrylamide gel	invertase	70	12	40	20	27
	aminoacylase	70	10	5	30	35, 98
	aspartase	50	10	40	45	50

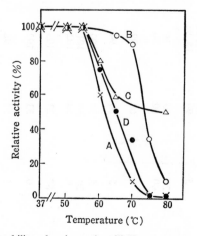

Fig. 3.10   Heat stability of aminoacylase.[98] Heat treatment was carried out for 10 min.
A, Native aminoacylase; B, DEAE-Sephadex-aminoacylase; C, iodoacetyl-cellulose-aminoacylase; D, polyacrylamide-aminoacylase.

have also been reported. For instance, the remaining activity of invertase immobilized by ionic binding with DEAE-cellulose after treatment at 50°C for 30 min was 40%, while the native enzyme retained 100% of its initial activity after similar treatment.[67]

As described above, enhancement of heat stability on immobilization has been observed in many enzymes, but no correlation between heat stability and immobilization method has yet been established. However, most enzymes immobilized by physical adsorption showed a decrease of heat stability. Therefore, this immobilization method is considered to be unsuitable where high heat stability of the enzyme is required.

The authors have investigated the heat stabilities of immobilized microbial cells. The remaining aspartase activity of intact *Escherichia coli* cells after treatment at 55°C for 30 min was 10%, while in immobilized *E. coli* cells prepared by entrapping in polyacrylamide gel, 40% of the initial activity was retained after similar treatment.[53] Similar results were obtained with arginine deiminase activity of *Pseudomonas putida* cells[56] and penicillin amidase activity of *E. coli* cells[57] immobilized by the same method. On the other hand, the heat stability of fumarase activity of *Brevibacterium ammoniagenes* cells did not change on immobilization.[54]

3.5.4   Storage Stability

The storage stability is an important factor to be considered in appli-

TABLE 3.11 Storage stabilities of immobilized enzyme

Enzyme	Immobilization method	Storage		Remaining activity		Reference
		Temp. (°C)	Period (days)	Native enzyme	Immobilized enzyme	
Lactate dehydrogenase	carrier cross-linking (porous glass + glutaraldehyde)	25	30	0	100	91
Glyceraldehyde-phosphate dehydrogenase	carrier cross-linking (AE-cellulose + glutaraldehyde)	23	1	0	95	39
Acetylcholinesterase	peptide binding (butanediol-divinyl ether-maleic anhydride copolymer)	4	140	70	100	105
Esterase	peptide binding (porous glass azide)	4 25	14 14	100 100	95 50	106 106
Naringinase	peptide binding (isobutylvinyl ether-maleic anhydride copolymer)	4	7	90	50	107
	(ethylene-maleic anhydride copolymer)	4	7	90	100	107
	(styrene-maleic anhydride copolymer)					
	(methylvinyl ether-maliec anhydride copolymer)	4	7	90	50	107
		4	7	90	100	107
Asparaginase	entrapping (polyurea microcapsules)	4	20	50	70	44
Urease	entrapping (cellulose acetate butyrate)	22 38	15 15	47 10	100 42	108 108
Aldolase	carrier cross-linking (AE-cellulose + glutaraldehyde)	23	21	70	95	39

cations of immobilized enzymes, and many reports have been published, as shown in Table 3.11.

An increase of storage stability on immobilization of enzymes was observed with lactate dehydrogenase immobilized by glutaraldehyde on porous glass,[91] and with trypsin[95] and chymotrypsin[46] immobilized by diazo binding to amino acid copolymer. The immobilized trypsin could be stored for several months without loss of activity at 2°C in 0.0025 N hydrochloric acid. This stabilization was considered to be caused by a reduction of autolysis, i.e., blocking of the catalytic activity of trypsin toward substrate lysine residue, since the ε-amino group of lysine in the enzyme protein was protected by immobilization.

The authors investigated the storage stability of aminoacylase immobilized by ionic binding with DEAE-Sephadex in 0.2 M acetyl-DL-methionine solution and in distilled water at various temperatures.[15] It was found that 60% of the initial activity was retained after storage for 150 days at 4°C in both solutions, while the remaining activity, after storage for 150 days at 25°C was 50% in acetyl-DL-methionine solution and 10% in distilled water.

On the other hand, decreases of storage stability on immobilization have been reported. The remaining activity of naringinase immobilized by covalent binding with styrene-maleic anhydride copolymer was 50% of the initial activity after storage for a week at 4°C, while no decrease of activity was observed in the native enzyme.[107]

Further, the storage stabilities of trypsin, papain, and ficin immobilized by various methods using different carriers were studied.[109] It was found that inorganic materials such as porous glass were superior to organic materials such as cellulose as carriers for immobilization of these enzymes as regards storage stability. Of the binding methods to the inorganic carrier, diazo binding was superior as regards storage stability. However, very stable immobilized enzymes bound to organic carriers have been obtained, and it cannot be said generally that inorganic supports are superior to organic supports.

In relation to the storage stability of immobilized microbial cells, the authors observed that aspartase of *Escherichia coli*, entrapped in polyacrylamide gel retained 50% of the initial activity after storage for 4 months at 25°C or 6 months at 4°C in 1 M ammonium fumarate solution.

3.5.5 Operational Stability

A. Stability of immobilized enzymes

The operational stability of activity of immobilized enzymes is one of the most important factors affecting the success of industrialization of

TABLE 3.12 Operational stabilities of immobilized enzymes

Enzyme	Immobilization method	Application	Operation		Remaining	Reference
			Temp. (°C)	Period (days)	activity (%)	
β-Tyrosinase	covalent binding (Sepharose-coenzyme)	synthesis of L-tyrosine	25	7	70	110
Phosphorylase	carrier cross-linking (porous glass+ glutaraldehyde)	synthesis of poly-saccharide	45	28	50	111
Carbamate kinase	carrier cross-linking (porous glass+ glutaraldehyde)	regeneration of ATP	40	15	74	74
β-Amylase	peptide binding (acrylamide-acrylic acid copolymer+ carbodiimide)	hydrolysis of starch	40	20	50	112
Glucoamylase	ionic binding (DEAE-cellulose)	saccharification of starch	40	21	100	113, 114
β-Galactosidase	carrier cross-linking (ZrO₂-coated porous glass+glutaraldehyde)	hydrolysis of sweet whey hydrolysis of lactose	40 30	54 100	50 50	115 115
Invertase	diazo binding (porous glass) entrapping (cellulose triacetate fibers)	inversion of sucrose inversion of sucrose	23 25 25	28 1000 5300	100 100 50	79 116 117

(*Continued*)

TABLE 3.12—*Continued*

Enzyme	Immobilization method	Application	Operation Temp. (°C)	Operation Period (days)	Remaining activity (%)	Reference
Pepsin	peptide binding (porous glass + carbodiimide)	hydrolysis of hemo-globin	23	25	100	118
Papain	carrier cross-linking (porous glass + glutaraldehyde)	hydrolysis of casein	45	35	50	32
Urease	diazo binding (porous glass)	decomposition of urea	25	8	90	119
Penicillin amidase	alkylation (dichloro-s-triazinyl-cellulose)	production of 6-APA	37	77	100	120
Aminoacylase	ionic binding (DEAE–Sephadex)	production of L-amino acid	50	30	60–70	98
	carrier cross-linking (porous glass + glutaraldehyde)	production of L-amino acid	37	78	50	121
	entrapping (polyacrylamide gel)	production of L-amino acid	37	20	90	35
AMP deaminase	ionic binding (DEAE-cellulose)	production of 5'-IMP	55	5	70	81
Aspartase	entrapping (polyacrylamide gel)	production of L-aspartic acid	37	20	50	50
Glucose isomerase	diazo binding (porous glass)	isomerization of glucose	60	14	50	62

an immobilized system. Thus, many reports have been published on operational stability, as listed in Table 3.12.

A column of immobilized invertase prepared by diazo binding with porous glass still retained full initial activity after 28 days of continuous operation at 23°C with 20% sucrose solution.[79] The remaining activity of a column of potato phosphorylase prepared by means of glutaraldehyde on porous glass was 50% after 28 days of continuous operation at 45°C with 0.01 M D-glucosylphosphate.[111] In the case of a column of immobilized invertase prepared by entrapping in cellulose triacetate fibers, the column was reported to retain its initial activity even after 3 years of continuous hydrolysis of sucrose at 25°C.[116] Further, in the case of polynucleotide phosphorylase immobilized on CNBr-activated cellulose, repeated use up to 40 times was possible.[122] Glucose oxidase and urease immobilized by means of glutaraldehyde on nylon tube could be used for automatic determination of 150 samples daily of glucose and urea over 30 days.[123]

The authors observed that no loss of activity of asparaginase microencapsulated by nylon or polyurea membrane occurred after repeated use up to 5 times, indicating that asparaginase was completely entrapped into microcapsules.[44]

B. Stability of immobilized microbial cells

The stability of immobilized microbial cells in continuous operation or upon repeated use has also been studied.

A column of immobilized lichens prepared by entrapping in polyacrylamide gel retained esterase and decarboxylase activities after 3 weeks of operation at 30°C.[124]

The authors compared the operational stabilities of immobilized and intact cells for aspartase activity of *Escherichia coli* and fumarase activity of *Brevibacterium ammoniagenes*. As shown in Fig. 3.11, the activities of both microbial cells were markedly stabilized by immobilization. Thus, it is apparent that immobilized microbial cells are advantageous for practical use. The operational stabilities of various immobilized microbial cells under specific conditions are listed in Table 3.13. As is clearly shown in the table, immobilized microbial cells are very stable compared with intact cells. More recently, the authors[133] found that aspartase activity of *E. coli* treated with glutaraldehyde and hexamethylenediamine after immobilization of the cells using κ-carrageenan is more stable than that of immobilized cells prepared by polyacrylamide gel method.

In order to investigate the stabilization mechanism, the authors carried out some experiments on the aspartase activity of *E. coli* cells. At first, we considered that aspartase of immobilized microbial cells might be

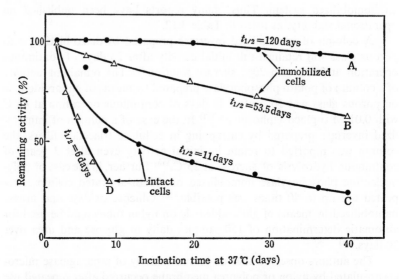

Fig. 3.11 Stability of enzyme activities of intact and immobilized cells.
A, aspartase activity of immobilized cells; B, fumarase activity of immobilized cells; C, aspartase activity of intact cells; D, fumarase activity of intact
cells. $t_{1/2}$, half life.

stabilized due to binding to some cellular granules or the cell membrane.
We therefore carried out the following experiments. That is, after sonica-
tion or autolysis of intact *E. coli* cells, a precipitate and soluble fraction
were obtained by centrifugation at 27,000 g for 20 min. The fractions were
immobilized separately by the polyacrylamide gel method, and the stabilities of these preparations were compared. As shown in Table 3.14, the
activity of the immobilized precipitate fraction was found to be very stable,
whereas that of the immobilized soluble fraction was very unstable.
Further, intact cells were preliminarily treated with deoxycholate or
Triton X-100 as a solubilizing agent for membrane-bound enzymes. On
centrifugation at 27,000 g for 20 min, the precipitating and soluble fractions were separated and immobilized separately by the polyacrylamide
gel method. The stabilities of these immobilized preparations were compared. As shown in Table 3.14, both immobilized preparations were unstable. These results indicate that aspartase becomes unstable upon solu-
bilization. In other words, aspartase may be stabilized by binding with
cellular granules or cell membranes.

In order to confirm this assumption, the authors carried out model
experiments. Native asparatase was separately immobilized by covalent
binding, ionic binding, and hydrophobic binding with water-insoluble
carriers. Then, these immobilized preparations were further entrapped in

TABLE 3.13 Conditions for continuous production of organic compounds by immobilized microbial cells

Enzyme	Aspartase	Fumarase	L-Arginine deiminase	L-Histidine ammonia-lyase	Penicillin amidase
Microorganism	*E. coli*	*B. ammoniagenes*	*P. putida*	*A. liquidum*	*E. coli*
Substrate	1.0 M NH$_4$-fumarate	1.0 M Na-fumarate	0.5 M L-arginine	0.25 M L-histidine	0.05 M penicillin G
pH	8.5	7.0	6.0	9.0	8.5
Stabilizer	Ca^{2+}, Mg^{2+}, or Mn^{2+}	—	—	Ca^{2+}, Co^{2+}, Mg^{2+} or Zn^{2+}	—
Temperature (°C)	37°	37°	37°	37°	40°
Flow rate† (SV, h^{-1})	0.80	0.23	0.26	0.06	0.24
Product	L-aspartic acid	L-malic acid	L-citrulline	urocanic acid	6-APA
Yield of product (%)	95	70	96	91	80
Half-life (days)	120	52.5	140	180	17 (at 40°C) 42
Reference	53, 125	54	56	55	57 (at 30°C)

† Data show flow rates for complete conversion of substrate to product.

TABLE 3.14   Stability of aspartase activity of various
immobilized preparations

Preparation	Half-life of activity (days)†
Cells (no pretreatment)	120
Sonicated cells	38
ppt fraction	76
Soluble fraction	6
Deoxycholate-treated cells	
Autolysate	9
ppt fraction	13
Soluble fraction	8
Immobilized aspartases	
Covalent binding to Sepharose	>30
Ionic binding to TEAE-cellulose	>30
Hydrophobic binding to aminopentyl-Sepharose	>30

† Continuous enzyme reaction

polyacrylamide gel, and the operational stabilities of the resulting immobilized preparations were investigated. As shown in Table 3.14, aspartase was found to be very stable when native aspartase was immobilized
using a water-insoluble carrier and then entrapped in polyacrylamide gel.
These results support the above assumption. Therefore, it seems clear that
in some cases the use of immobilized microbial cells is preferrable for
continuous enzyme reactions than that of immobilized enzyme.

3.6   OTHER PROPERTIES

In addition to the enzymatic properties described above, the effect of
pH on the stability of immobilized enzymes has been studied. The stabilities of trypsin,[30,126] chymotrypsin,[126] and papain[126] were increased on
the alkaline side when they were immobilized using polyanionic carriers,
such as a copolymer of styrene and maleic anhydride, while the stability
of enzymes immobilized using polycationic carriers was enhanced on the
acid side. The pH dependence of the stability of immobilized papains on
neutral carriers such as p-aminobenzyl cellulose,[126,127] leucine-p-amino-
phenylalanine copolymers,[128] and starch-methylenedianiline resins[126,127]
was similar to that of the native enzyme (refer to 2.1.1. C). However, immobilized subtilisin Carlsberg prepared with the same carriers showed
increased stability in the acid pH region (pH 3–4).[126,127] The water-soluble
chemically modified polytyrosyl trypsin was immobilized using the same
carriers, and it exhibited maximal stability in the neutral and alkaline

pH regions, in contrast to native trypsin or polytyrosyl trypsin, which are most stable in the acid pH region.[40,41,129]

Improved stability toward oxidation was reported in bromelain[130] and ficin[131] immobilized by peptide binding with CM-cellulose azide.

The membrane permeability of substrate and product and the possibility of side reactions are the important current areas of investigation with immobilized microbial cells. The authors immobilized *Achromobacter liquidum* cells having histidine ammonia-lyase activity[55] and *Pseudomonas putida* cells having arginine deiminase activity[56] by the polyacrylamide gel method, and investigated the application of these immobilized microbial cells for the continuous production of urocanic acid and L-citrulline, respectively. Intact cells of these microorganisms scarcely exhibit the desired catalytic activity, and the enzyme reaction does not proceed due to the low permeability of substrate and/or product through the cell membrane. However, as shown in Fig. 3.12, the apparent catalytic activity is enhanced, and the enzyme reaction proceeds upon the addition of cationic detergents such as cetyltrimethyl ammonium bromide (CTAB) to the substrate solution. On the other hand, immobilized microbial cells can catalyze the reaction without the addition of CTAB. This indicates that some structural changes occur in the cell membrane upon immobilization, and the selectivity toward substrate and/or product is lost or reduced. That is, the membrane permeability barrier is removed. This is one of the advantages of using immobilized microbial cells.

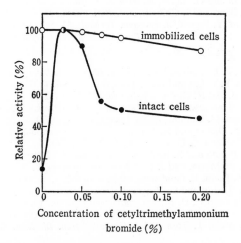

Fig. 3.12 Effect of cetyltrimethylammonium bromide on the formation of urocanic acid by *Achromobacter liquidum* cells.
Reaction temperature, 37°C; reaction time, 60 min.

Spores of *Aspergillus oryzae* were immobilized by using CM-cellulose, P-cellulose, DEAE-cellulose or ECTEOLA-cellulose, and the invertase activity of the immobilized spore column was measured.[132] The reaction rate decreased somewhat when the column length increased above a certain range. It may be considered that the reaction product is subsequently decomposed by other contaminating enzymes in the column.

Thus, in the case of immobilized microbial cells, it is necessary to consider side reactions catalyzed by other enzymes existing in the system.

REFERENCES

1. H. Maeda and H. Suzuki, *Nippon Nogeikagaku Kaishi*, **44**, 547 (1970)
2. T. Takami and T. Ando, *Seikagaku*, **40**, 749 (1968)
3. R. Axén, P. Vretblad, and J. Porath, *Acta Chem. Scand.*, **25**, 1129 (1971)
4. R. Kleine, P. Spangenberg, and C. Flemming, *Hoppe-Seyler's Z. Physiol. Chem.*, **357**, 629 (1976)
5. A. Patchornik, Israel Patent, 18207 (1965)
6. R. Haynes and K. A. Walsh, *Biochem. Biophys. Res. Commun.*, **36**, 235 (1969)
7. L. Sundberg and T. Kristiansen, *FEBS Letters*, **22**, 175 (1972)
8. J. J. Cebra, D. Givol, H. I. Silman, and E. Katchalski, *J. Biol. Chem.*, **236**, 1720 (1961)
9. E. F. Jansen and A. C. Olson, *Arch. Biochem. Biophys.*, **129**, 221 (1969)
10. W. E. Hornby, M. D. Lilly and E. M. Crook, *Biochem. J.* **98**, 420 (1966)
11. G. P. Royer and G. M. Green, *Biochem. Biophys. Res. Commun.*, **44**, 426 (1971)
12. P. Cresswell and A. R. Sanderson, *Biochem. J.*, **119**, 447 (1970)
13. T. Tosa, T. Mori, N. Fuse, and I. Chibata, *Enzymologia*, **32**, 153 (1967)
14. T. Tosa, T. Mori, N. Fuse, and I. Chibata, *Biotechnol. Bioeng.*, **9**, 603 (1967)
15. T. Tosa, T. Mori and I. Chibata, *Agr. Biol. Chem.*, **33**, 1053 (1969)
16. H. H. Weetall and G. Baum, *Biotechnol. Bioeng.*, **12**, 339 (1970)
17. T. Tosa, R. Sano and I. Chibata, *Agr. Biol. Chem.*, **38**, 1529 (1974)
18. K. Kawai and Y. Eguchi, *J. Ferment. Technol.*, **53**, 588 (1975)
19. O. R. Zaborsky and J. Ogletree, *Biochim. Biophys. Acta*, **289**, 68 (1972)
20. J. C. Smith, I. J. Statford, D. W. Hutchinson and H. J. Brentnall, *FEBS Letters*, **30**, 246 (1973)
21. S. P. O'neill, P. Dunnill and M. D. Lilly, *Biotechnol. Bioeng.*, **13**, 337 (1971)
22. D. R. Marsh, Y. Y. Lee and G. T. Tsao, *ibid.*, **15**, 483 (1973)
23. D. T. Wadiak and R. G. Carbonell, *ibid.*, **17**, 1157 (1975)
24. S. Usami, E. Hasegawa and M. Karasawa, *Hakko Kyokaishi*, **33**, 152 (1975)
25. P. Monsan and G. Durand, *FEBS Letters*, **16**, 39 (1971)
26. H. Suzuki, Y. Ozawa and H. Maeda, *Agr. Biol. Chem.*, **30**, 807 (1966)
27. K. Kawashima and K. Umeda, *ibid.*, **40**, 1157 (1976)
28. R. Axén, J. Porath and S. Ernback, *Nature*, **214**, 1302 (1967)
29. R. Axén, P. A. Myrin, and J. C. Janson, *Biopolymers*, **9**, 401 (1970)
30. Y. Levin, M. Pecht, L. Goldstein and E. Katchalski, *Biochemistry*, **3**, 1905 (1964)
31. R. Uy, V. S. H. Liu and G. P. Royer, *J. Solid-Phase Biochem.*, **1**, 51 (1976)
32. H. H. Weetall and R. D. Mason, *Biotechnol. Bioeng.*, **15**, 455 (1973)
33. T. Mori, T. Tosa and I. Chibata, *Cancer Res.*, **34**, 3066 (1974)
34. T. Sato, T. Mori, T. Tosa and I. Chibata, *Arch. Biochem. Biophys.*, **147**, 788 (1971)
35. T. Mori, T. Sato, T. Tosa and I. Chibata, *Enzymologia*, **43**, 213 (1972)
36. H. Maskova, T. Barth, B. Jirovsky and Z. Rychlik, *Collection Czech. Chem. Commun,*, **38**, 943 (1973)

37. S. Chung, M. Hamano, K. Aida and T. Uemura, *Agr. Biol. Chem.*, **32**, 1287 (1968)
38. A. B. Patel, S. N. Pennington and H. D. Brown, *Biochim. Biophys. Acta*, **178**, 626 (1969)
39. R. D. Falb, J. Lynn and J. Shapira, *Experientia*, **28**, 958 (1973)
40. L. Goldstein and E. Katchalski, *Z. Anal. Chem.*, **243**, 375 (1968)
41. L. Goldstein, Y. Levin and E. Katchalski, *Biochemistry*, **3**, 1913 (1964)
42. P. Bernfeld, R. E. Bieber and D. M. Watson, *Biochim. Biophys. Acta*, **191**, 570 (1969)
43. H. Filippusson and W. E. Hornby, *Biochem. J.*, **120**, 215 (1970)
44. T. Mori, T. Tosa and I. Chibata, *Biochim. Biophys. Acta*, **321**, 653 (1973)
45. P. Bernfeld and R. E. Bieber, *Arch. Biochem. Biophys.*, **131**, 587 (1969)
46. E. Katchalski, *Polyamino Acids, Polypeptides, and Proteins*, p. 283, The Univ. of Wisconsin Press, 1962
47. A. Bar-Eli and E. Katchalski, *J. Biol. Chem.*, **238**, 1690 (1963)
48. R. Axén, J. Carlsson, J. C. Janson and J. Porath, *Enzymologia*, **41**, 359 (1971)
49. K. S. Stenn and E. R. Blout, *Biochemistry*, **11**, 4502 (1972)
50. T. Tosa, T. Sato, T. Mori, Y. Matuo and I. Chibata, *Biotechnol. Bioeng.*, **15**, 69 (1973)
51. T. Hattori and C. Furusaka, *J. Biochem.*, **48**, 831 (1960)
52. T. Hattori and C. Furusaka, *ibid.*, **50**, 312 (1961)
53. I. Chibata, T. Tosa and T. Sato, *Appl. Microbiol.*, **27**, 878 (1974)
54. K. Yamamoto, T. Tosa, K. Yamashita and I. Chibata, *European J. Appl. Microbiol.*, **3**, 169 (1976)
55. K. Yamamoto, T. Sato, T. Tosa and I. Chibata, *Biotechnol. Bioeng.*, **16**, 1601 (1974)
56. K. Yamamoto, T. Sato, T. Tosa and I. Chibata, *ibid.*, **16**, 1589 (1974)
57. T. Sato, T. Tosa and I. Chibata, *European J. Appl. Microbiol.*, **2**, 153 (1976)
58. W. Brümmer, N. Hennrich, M. Klockow, H. Lang and H. D. Orth, *Eur. J. Biochem.*, **25**, 129 (1972)
59. S. Ikeda and S. Fukui, *Biochem. Biophys. Res. Commun.*, **52**, 482 (1973)
60. W. H. Bishop, F. A. Quiocho, and F. M. Richards, *Biochemistry*, **5**, 4077 (1966)
61. K. Venkatasubramanian, W. R. Vieth and S. S. Wang, *J. Ferment. Technol.*, **50**, 600 (1972)
62. G. W. Strandberg and K. L. Smiley, *Biotechnol. Bioeng.*, **14**, 509 (1972)
63. G. P. Royer and J. P. Andrews, *J. Biol. Chem.*, **248**, 1807 (1973)
64. H. P. Gregor, and P. W. H. Rauf, *Biotechnol. Bioeng.*, **17**, 445 (1975)
65. B. Walter, *Biochim. Biophys. Acta*, **429**, 950 (1976)
66. A. Ya. Nikolaev, *Biochemistry*, (USSR) (English Transl.), **27**, 713 (1962)
67. A. Usami, J. Noda and K. Goto, *J. Ferment. Technol.*, **49**, 598 (1971)
68. A. K. Sharp, G. Kay and M. D. Lilly, *Biotechnol. Bioeng.*, **11**, 363 (1969)
69. Y. Kosugi and H. Suzuki, *J. Ferment. Technol.*, **51**, 895 (1973)
70. J. E. Dixon, F. E. Stolzenbach, *Israel J. Chem.*, **12**, 529 (1974)
71. H. H. Weetall, *Res. Develop.*, **22**, No. 12, 18 (1971)
72. W. E. Hornby, M. D. Lilly and E. M. Crook, *Biochem. J.*, **107**, 669 (1968)
73. H. Maeda and H. Suzuki, *Agr. Biol. Chem.*, **36**, 1581 (1972)
74. F. Widmer, J. E. Dixon and N. O. Kaplan, *Anal. Biochem.*, **55**, 282 (1973)
75. S. W. May and N. N. Li, *Biochem. Biophys. Res. Commun.*, **47**, 1179 (1972)
76. K. P. Wheeler, B. A. Edwards and R. Whittam, *Biochim. Biophys. Acta*, **191**, 187 (1969)
77. P. A. Srere, B. Mattiasson and K. Mosbach, *Proc. Nat. Acad. Sci.*, **70**, 2534 (1973)
78. R. C. Boguslaski and A. M. Janik, *Biochim. Biophys. Acta*, **250**, 266 (1971)
79. R. D. Mason and H. H. Weetall, *Biotechnol. Bioeng.*, **14**, 637 (1972)
80. B. Svensson, *Compt. Rend. Trav. Lab. Carlsberg*, **39**, 1 (1972)
81. H. Filippusson and W. E. Hornby, *Biochem. J.*, **120**, 215 (1970)

82. S. Usami and Y. Kuratsu, *J. Ferment. Technol.*, **51**, 789 (1973)
83. G. J. H. Melrose, *Rev. Pure Appl. Chem.*, **21**, 83 (1971)
84. S. N. Pennington, H. D. Brown, A. B. Patel and C. O. Knowles, *Biochim. Biophys. Acta*, **167**, 479 (1968)
85. T. Horigome, H. Kasai and T. Okuyama, *J. Biochem.*, **75**, 299 (1974)
86. S. Usami, T. Yamada and A. Kimura, *Hakko Kyokaishi*, **25**, 513 (1967)
87. C. Schwabe, *Biochemistry*, **8**, 795 (1969)
88. D. Gabel, P. Vretblad, R. Axén and J. Porath, *Biochim. Biophys. Acta*, **214**, 561 (1970)
89. C. K. Glassmeyer and J. D. Ogle, *Biochemistry*, **10**, 786 (1971)
90. P. L. Osheroff and R. J. Guillory, *Biochem. J.*, **127**, 419 (1972)
91. J. E. Dixon, F. E. Stolzenbach, J. A. Berenson and N. O. Kaplan, *Biochem. Biophys. Res. Commun.*, **52**, 905 (1973)
92. A. Schejter and A. Bar-Eli, *Arch. Biochem. Biophys.*, **136**, 325 (1970)
93. A. E. Chung, *ibid.*, **152**, 125 (1972)
94. S. Litvak, L. T. Litvak, D. S. Carre and F. Chapeville, *Eur. J. Biochem.*, **24**, 249 (1971)
95. E. Katchalski and A. Bar-Eli, *Nature*, **188**, 856 (1960)
96. H. Sekine, *Agr. Biol. Chem.*, **37**, 437 (1973)
97. T. Mori, T. Sato, Y. Matuo, T. Tosa and I. Chibata, *Biotechnol. Bioeng.*, **14**, 663 (1972)
98. I. Chibata, T. Tosa, T. Sato, T. Mori and Y. Matsuo, *Proc. IV I.F.S.: Ferment. Technol. Today*, p. 383, Society of Fermentation Technology, Japan, 1972
99. M. A. Goheer, B. J. Gould and D. V. Parke, *Biochem. J.*, **157**, 289 (1976)
100. T. Toraya, K. Ohashi and S. Fukui, *Biochemistry*, **14**, 4255 (1975)
101. N. W. H. Cheethan and G. N. Richards, *Carbohydrate Res.*, **30**, 99 (1973)
102. A. B. Patel, R. O. Stasiw, H. D. Brown and C. A. Ghiron, *Biotechnol. Bioeng.*, **14**, 1031 (1972)
103. Y. Ohno and M. Stahmann, *Macromolecules*, **4**, 350 (1971)
104. P. D. Weston and S. Avrameas, *Biochem. Biophys. Res. Commun.*, **45**, 1574 (1971)
105. C. Alsen, U. Bertram, T. Gersteuer and F. K. Ohnesorge, *Biochim. Biophys. Acta*, **377**, 297 (1975)
106. M. K. Weibel, R. Barrios, R. Delotto and A. E. Humphrey, *Biotechnol. Bioeng.*, **17**, 85 (1975)
107. L. Goldstein, A. Lifshitz and M. Sokolvsky, *Int. J. Biochem.*, **2**, 448 (1971)
108. D. L. Gardner, R. D. Falb, B. C. Kim and D. C. Emmerling, *Trans. Am. Soc. Artificial Internal Organs*, **17**, 239 (1971)
109. H. H. Weetall, *Biochim. Biophys. Acta*, **212**, 1 (1970)
110. S. Fukui, S. Ikeda, M. Fujimura, H. Yamada and H. Kumagai, *Eur. J. Appl. Microbiol.*, **1**, 25 (1975)
111. D. L. Marshall and J. L. Walter, *Carbohydrate Res.*, **25**, 489 (1972)
112. K. Martensson, *Biotechnol. Bioeng.*, **16**, 1567 (1974)
113. M. J. Bachler, G. W. Strandberg and K. L. Smiley, *ibid.*, **12**, 85 (1970)
114. K. L. Smiley, *ibid.*, **13**, 309 (1971)
115. H. H. Weetall, N. B. Havewala, W. H. Pitcher, C. C. Detar, W. P. Vann and S. Yaverbaum, *ibid.*, **16**, 689 (1974)
116. D. Dinelli, *Process Biochem.*, **7**, 9 (1972)
117. W. Marconi, S. Gulinelli and F. Morisi, *Biotechnol. Bioeng.*, **16**, 501 (1974)
118. W. F. Line, A. Kwong and H. H. Weetall, *Biochim. Biophys. Acta*, **242**, 194 (1971)
119. K. B. Ramachandran and D. D. Perlmutter, *Biotechnol. Bioeng.*, **18**, 685 (1976)
120. D. A. Self, G. Kay, M. D. Lilly and P. Dunnill, *ibid.*, **11**, 337 (1969)
121. Y. Yokote, M. Fujita, G. Shimura, S. Noguchi, K. Kimura and H. Samejima, *J. Solid-Phase Biochem.*, **1**, 1 (1976)
122. C. H. Hoffman, E. Harris, S. Chodroff, S. Michelson, J. W. Rothrock, E. Peterson and W. Reuter, *Biochem. Biophys. Res. Commun.*, **41**, 710 (1970)

123. D. J. Inman and W. E. Hornby, *Biochem. J.*, **129**, 255 (1972)
124. K. Mosbach and R. Mosbach, *Acta Chem. Scand.*, **20**, 2807 (1966)
125. T. Tosa, T. Sato, Y. Nishida and I. Chibata, *Biochim. Biophys. Acta*, **483**, 193 (1977)
126. L. Goldstein, *Method in Enzymology*, vol. 19, p. 935, Academic Press, 1970
127. L. Goldstein, M. Pecht, S. Blumberg, D. Atlas and Y. Levin, *Biochemistry*, **9**, 2322 (1970)
128. I. H. Silman, M. Albu-Weissenberg and E. Katchalski, *Biopolymers*, **4**, 441 (1966)
129. L. Goldstein, *Fermentation Advances* (ed. D. Perlman), p. 391, Academic Press, 1969
130. C. W. Wharton, E. M. Crook and K. Brocklehurst, *Eur. J. Biochem.*, **6**, 565 (1968)
131. M. D. Lilly, C. Money, W. E. Hornby and E. M. Crook, *Biochem. J.*, **95**, 45p (1965)
132. D. E. Johnson and A. Ciegler, *Arch. Biochem. Biophys.*, **130**, 384 (1969)
133. T. Tosa, T. Sato, K. Yamamoto, I. Takata, Y. Nishida and I. Chibata, 26th International Congress of Pure and Applied Chemistry, at Tokyo, p. 267, September 4, 1977

CHAPTER 4

Applications of immobilized enzymes and microbial cells

Because of the special characteristics of immobilized enzymes, as described in the previous chapter, a considerable amount of work has been carried out on applications of immobilized enzymes as solid catalysts having high substrate specificity, and more recently, interest has also extended to immobilized microbial cells.

Immobilized enzymes and immobilized microbial cells are used in various fields, such as chemical processes, analysis, medical treatment, food processing, affinity chromatography, and so forth. For the efficient utilization of such immobilized systems, chemical engineering studies are necessary.

4.1 CHEMICAL ENGINEERING OF IMMOBILIZED SYSTEMS

Among the applications of immobilized enzymes and immobilized microbial cells, utilization for chemical processes is the most active and important field.

In the production of chemicals by the action of enzymes, it should be advantageous compared with the conventional enzyme reaction using native enzyme if stable immobilized enzymes can be used, because even in a batch method, the immobilized enzymes can be recovered after enzyme reaction by centrifugation or filtration, and the recovered immobilized enzymes can be reused. In such an immobilized system, the isolation and purification of reaction products can easily be carried out. Further, if the immobilized enzymes are packed into a column, a continuous enzyme reaction becomes possible. Thus, an immobilized system has distinct advantages for industrial use, readily permitting automatic operation, increasing the yield of product per unit of enzyme, simplifying the product purification procedure, and so forth.

For these reasons, studies on the use of immobilized enzymes and immobilized microbial cells for chemical processes have been developing.

In this section, the basic chemical engineering considerations for application of immobilized systems to chemical processes are described.

4.1.1 Enzyme Reactors

Lilly and Dunnill[1] classified biochemical reactions using enzymes or microbial cells as shown in Table 4.1, according to the type of catalyst and the mode of operation.

TABLE 4.1 Biochemical reactors[1]

Kind of catalyst			Type of reactor		
Microbial cells	metabolizing (fermentation) non-metabolizing (resting) immobilized	batch	single use reuse		
Enzyme	soluble soluble immobilized insoluble immobilized	continuous	open	tubular stirred	
			closed	tubular stirred	

There are two groups of catalysts, i.e. microbial cells and the enzymes isolated or extracted from cells. In the case of microbial cells, metabolizing cells (as in "fermentation"), non-metabolizing cells (called "resting cells") or immobilized cells can be used. In the case of enzymes, the following three types can be considered; soluble forms, soluble immobilized forms used with ultrafiltration membranes, and insoluble immobilized forms. As for the type of enzyme reactor, batch and continuous systems can be employed.

These enzyme reactors can be classified by the mode of operation and the flow characteristics of substrate and product as shown schematically in Fig. 4.1. These reactors can also be classified as shown in Table 4.2, according to the mode of operation, the flow pattern and the type of reactor.

The characteristics of these reactors are described in the following sections. The choice of reactor must be made according to the intended application.

A. Stirred tank reactor

A batch stirred tank reactor (1A in Fig. 4.1) is the simplest type, and is composed of a reactor and a stirrer. As a stirrer, a turbine wing and a propeller are used. Obstacle boards are fitted to the tank sides to improve the efficiency of stirring.

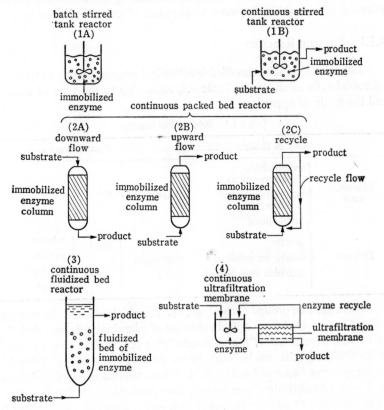

Fig. 4.1   Flow patterns for enzyme reactors using immobilized enzymes.[5]

TABLE 4.2   Classification of enzyme reactors

Mode of operation	Flow pattern	Type of reactor
Batch system	well-mixed	stirred tank reactor
Continuous system	well-mixed	single CSTR[†] multiple CSTR[†]
	plug flow	packed bed fluidized bed enzyme tube enzyme film hollow fiber

† CSTR: continuous stirred tank reactor

This reactor is useful for substrate solutions of high viscosity and for immobilized enzymes relatively low activity. However, there is the difficulty that an immobilized enzyme tends to decompose on physical stirring.

The batch system is generally suitable for the production of rather small amounts of chemicals.

A continuously stirred tank reactor (1B in Fig. 4.1) is more efficient than a batch stirred tank reactor, but the reactor equipment is slightly more complicated.

B. Packed bed reactor

Continuous packed bed reactors (2A, 2B and 2C in Fig. 4.1) are the most widely used reactors for immobilized enzymes and immobilized microbial cells. In these cases, it is necessary to consider the pressure drop across the packed bed, i.e., column, and the effect of column dimensions on the reaction rate.

As regards substrate flow, there are three possibilities: i) downward flow method (2A), ii) upward flow method (2B), and iii) recycling method (2C). When the linear velocity of substrate solution affects the reaction rate, the recycling method is advantageous, as substrate solution can be passed through the column at a desired velocity.

For industrial applications, the flow direction of substrate solution is important. In the case of immobilized aminoacylase, the reaction rates for an aminoacylase column are equal with upward and downward flows.[2] However, in some cases downward flow causes compression of the beds of enzyme columns, so upward flow is generally preferred for industrial applications. Further, when gas is produced during an enzyme reaction, upward flow is again preferred.

The continuous method has the following advantages compared with batch methods: i) easy automatic control and operation, ii) reduction of labour costs, iii) stabilization of operating conditions, and iv) easy quality control of products.

C. Fluidized bed reactor

When a substrate solution of high viscosity and a gaseous substrate or product are used in a continuous reaction system, a continuous fluidized bed reaction (3 in Fig. 4.1) is most suitable. In this system, care must be taken to avoid destruction and decomposition of immobilized enzymes. The particle size of immobilized enzymes is important for the formation of a smooth fluidized bed. Recently, a tapered fluidized bed using immobilized enzyme has also been reported.[3,4]

D. Ultrafiltration membrane devices

A continuous ultrafiltration membrane device (4 in Fig. 4.1) is suitable for a substrate of high molecular weight and a product of low molecular weight. As the enzyme used in this case is soluble, no improvement of the stability of the enzyme can be expected. A hollow fiber device can also be used, and its characteristics are essentially the same as those of an ultra-filtration membrane.

E. Enzyme tubes

In most of the enzyme tubes that have been reported, the inner diameter was around 1 mm. Therefore, the Reynolds number is small, and the flow of substrate solution is laminar. However, when the inner surface of the tube is covered with gel, turbulent flow occurs due to the coarse inner surface.

F. Enzyme films

In order to increase the surface area of enzyme film per unit volume, various devices have been developed as practical equipment. For example, a spiral-type reactor was reported, as shown in Fig. 4.2.[5]

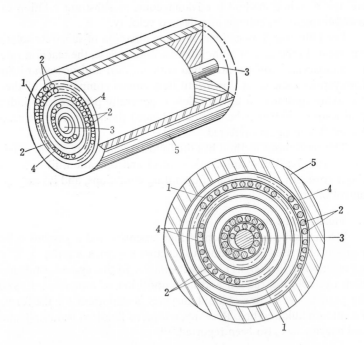

Fig. 4.2   Details of a multipore, spiral, biocatalytic module.[5]
Collagen-enzyme membrane (1) is coiled upon spacing elements (2). A rod

(*Continued overleaf*)

4.1.2   Pressure Drop

The pressure drop of an immobilized enzyme column is an important factor in continuous enzyme reactions using a packed bed reactor. To determine the pressure drop of a packed bed, the experimental equation (Eq. 4.1) presented by Kozeny[6] and Carman[7] for filtration can be applied.

$$\Delta P = k \frac{(1-\varepsilon)^2}{\varepsilon^3} \left[\frac{\phi}{D} \right]^2 \left[\frac{\mu u L}{gc} \right] \tag{4.1}$$

ΔP is the pressure drop, k is a constant, ε is the voidage, ϕ is the shape factor, D is the particle diameter, μ is the viscosity, u is the flow rate, L is the column length and gc is the gravity coefficient.

Putting $k \frac{(1-\varepsilon)^2}{\varepsilon^3} \left[\frac{\phi}{D} \right]^2 \left[\frac{\mu}{gc} \right]$

equal to K, Eq. 4.1 becomes

$$\Delta P = K u L \tag{4.2}$$

If K is a constant for a specific packed bed, the pressure drop of the column becomes proportional to the flow rate of substrate solution and the column length. In order to confirm whether Eq. 4.2 can be applied to enzyme columns, the authors[8] investigated the relationship between pressure drop, flow rate and column length using immobilized aminoacylase prepared by ionically binding the enzyme to DEAE-Sephadex. As shown in Fig. 4.3, it was confirmed that the Kozeny-Carman equation can be applied to an immobilized aminoacylase column. That is, the pressure drop is proportional to the flow rate and column length at a given temperature. The pressure drop also depends on temperature as well as the flow rate of substrate solution and column length. Thus, in Eq. 4.2, the term K containing the voidage, shape factor, particle diameter, and viscosity of substrate solution varies with temperature. Among these factors, viscosity seems to be most variable with temperature. Therefore, the pressure drop decreases with increase of temperature mainly due to a decrease of viscosity.

The authors[9] carried out similar experiments on *Escherichia coli* cells immobilized by the polyacrylamide gel method. As shown in Fig. 4.3, Eq. 4.2 can also be applied to a column packed with immobilized cells having a mean particle diameter of 3 mm. The pressure drop of this im-

---

(3) is used as a central spacer. The spacing elements are located at sufficiently narrow intervals to prevent the membrane from coming into contact with the layers above and below. The coiled enzyme-membrane cartridge contains multiple capillary compartments (4) which provide a larger surface area for efficient bulk mass interchange with the substrate. The cartridge is fitted into an outer shell (5) provided with an inlet and an outlet for the flow of substrate over the membrane surface.

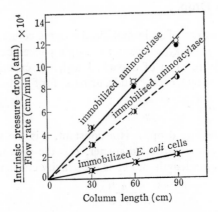

Fig. 4.3   Pressure drops of an immobilized DEAE-Sephadex aminoacylase column and an immobilized *E. coli* cell column.
A solution of 0.2 M acetyl-DL-methionine (pH 7.0, containing 5×10^{-4} M CoCl$_2$) was passed through various lengths of immobilized aminoacylase columns at the specified flow rate at 37°C or 50°C. A solution of 1 M ammonium fumarate (pH 8.5, containing 1×10^{-3} M MgCl$_2$) was passed through various lengths of immobilized *E. coli* cell columns at the specified flow rate at 37°C. The pressure drop was measured with a mercury manometer. Flow rate: ●—●, 2 cm/min; ○—○, 4 cm/min; ×—×, 8 cm/min. Temperature: ——, 37°C; ----, 50°C.

mobilized cell column is about 1/6 of that of a DEAE-Sephadex-aminoacylase column.

As shown in Eq. 4.1 the pressure drop depends on the particle diameter of immobilized enzymes and immobilized microbial cells. Therefore, when an ion-exchange resin of low cross-linkage level is used, it is necessary to consider changes in the degree of swelling due to ionic strength. Further, the pressure drop is markedly increased by a slight decrease of voidage. When the pressure drop increases, change of shape of immobilized enzymes or immobilized microbial cells may occur and the voidage becomes smaller, resulting in a further increase of the pressure drop.

In order to scale up an enzyme column for industrial purposes, the following factors must be suitably selected; appropriate materials for the column, shape of the column, and a flow pump suitable for the pressure drop.

4.1.3   Column Dimensions

To design an efficient packed bed reactor, the effect of column dimensions on the reaction rate should be investigated. In other words, it is necessary to know the effect of the linear velocity of substrate solution on the reaction rate.

The authors[2] tested the effect of column dimensions on the hydrolysis of acetyl-DL-methionine by using columns of various dimensions packed with immobilized aminoacylase. As shown in Table 4.3, essentially no significant difference of k value, the specific constant determining the rate of decomposition of the enzyme-substrate complex, was observed within the range of experimental conditions used (the ratio of column height and diameter was 1.5–21.4). Therefore, if the immobilized aminoacylase is uniformly packed into the column and the substrate solution is smoothly passed through the column, the enzyme reaction is considered to proceed at the same velocity regardless of the column dimensions. For industrial purposes, a flat and short column is considered to be advantageous from the viewpoint of pressure drop.

On the other hand, the shape of the column does influence on the reaction rate in some cases. For example, in the case of immobilized aspartase prepared by the polyacrylamide gel method, the authors[10] found that the velocity of formation of L-aspartic acid increased with increase of the column length at constant space velocity, as shown in Fig. 4.4. This indicates that the reaction rate increased due to a decrease of the external diffusion layer of the immobilized enzyme with increase of the linear velocity of the substrate solution.

However, even in this aspartase reaction, the formation of L-aspartic acid proceeded at the same rate regardless of the column dimensions in the case of aspartase activity of immobilized *Escherichia coli* cells.[11] The results are shown in Table 4.4. The rate of formation of L-aspartic acid was independent of the column dimensions over a range of height/diameter ratios from 4.27–14.94. The ratio of the diameter of immobilized cells to that of the column also had no effect on the reaction rate in the range of

TABLE 4.3 Effect of column dimensions on the rate constant
of decomposition of enzyme-substrate complex[†2]

Height of column (cm)	Diameter of column (cm)	Height Diameter	Rate constant, k (1/h)
3	2.00	1.5	59.1 ± 1.5
6	1.46	4.1	62.5 ± 0.9
6	0.84	7.1	61.7 ± 1.8
18	0.84	21.4	61.0 ± 0.3

† A solution of 0.2 M acetyl-DL-methionine (pH 7.0, containing 5×10^{-4} M Co^{2+}) was applied to the immobilized aminoacylase columns described in the table at the specified flow rates at 50°C.

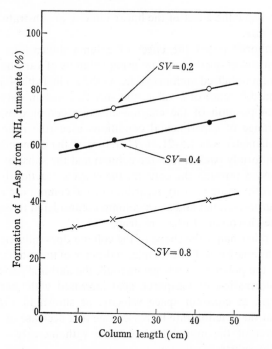

Fig. 4.4 Effect of column length on the formation of L-aspartic acid.[10] Using immobilized aspartase columns of various lengths (30 ml volume), a solution of 0.5 M NH₄-fumarate containing 1 mM Mn²⁺ was applied at 37°C at the indicated flow rates.

TABLE 4.4 Effect of column dimensions on reaction rate[11]

dp/D	H/D	Production of L-aspartic acid at (mol/liter)		
		$SV=1.0$	$SV=1.5$	$SV=2.0$
1/4	4.27	0.90	0.82	0.60
	8.23	0.89	0.81	0.58
	12.20	0.93	0.80	0.56
	14.94	0.89	0.80	0.56
1/12.5	8.00	0.92	—	0.53
1/25	10.00	0.90	—	—

† Immobilized *E. coli* cells (particle diameter, dp=4.0 mm) were packed into various columns; i.e., $D \times H$=16.4 × 70∼245 mm, 50 × 400 mm, and 100 × 1000 mm. The columns were maintained at 37°C, and a solution of 1 M ammonium fumarate (pH 8.5) containing 1 mM Mg²⁺ was passed through a column at the indicated flow rates. L-Aspartic acid produced in the effluent was measured.

1/4–1/25 under the flow conditions tested.

Therefore, it should be noted that the effect of column dimensions, i.e. the effect of the linear velocity of substrate solution on the reaction rate, differs depending on the state of the enzyme and the immobilization method even in the case of the same enzyme reaction.

4.1.4 Decay of Activity during Enzyme Reaction (stability)

The operational stability of immobilized enzymes and immobilized microbial cells is an important factor in the success or failure of industrialization of immobilized systems. This operational stability directly affects the costs of the enzyme and of repacking or regenerating deteriorated enzyme columns. Operational stability has already been discussed in section 3.5.5.

It is difficult to estimate the operational stability of immobilized enzymes and immobilized microbial cells from data on the storage and heat stabilities. Therefore, for the industrialization of immobilized systems it is important to investigate the changes of enzyme activity during long-term continuous operation by using a small test column.

A. Decay pattern of immobilized systems

The decay of enzyme columns packed with immobilized enzyme or immobilized microbial cells generally proceeds in an exponential pattern related to operating time. The patterns can be classified into 4 types as shown in Fig. 4.5.

Pattern [I] is a typical and normal one, and the decay proceeds at a constant rate. In pattern [II], the enzyme activity is apparently activated in the initial stage of operation, and then decay proceeds in the normal way. In the case of immobilized microbial cells in particular, an increase of apparent enzyme activity sometimes occurs in the intial stage of continuous enzyme reaction due to autolysis of cells, resulting in an increased passage of substrate and/or product. In the case of pattern [III], the enzyme activity is very stable for some time, but decreases rapidly from a certain stage for some reason. The following factors may be involved: i) incorrect operation, ii) bacterial contamination, iii) decrease of concentration of some stabilizing substance(s), and iv) accumulation of some inhibitory substance(s). In pattern [IV], the decrease of enzyme activity is rapid from the initial stage, but then the decay becomes very slow. In this case, immobilization of the enzyme is not complete, and leakage of enzyme from the immobilized preparation may occur in the initial stage. Further, in patterns [III] and [IV], it is possible that different enzyme proteins are immobilized and catalyze the same reaction.

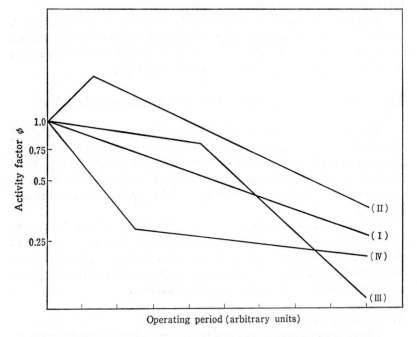

Fig. 4.5   Decay patterns of enzyme activity of immobilized systems.
I: Typical and normal pattern. II: Preliminary activation pattern. III: Two-phase pattern (stable→unstable). IV: Two-phase pattern (unstable→stable).

Factors influencing the decay of enzyme activity during continuous enzyme reactions can be summarized as follows.
1) Denaturation of the enzyme (due to heat of reaction, pH changes, and so forth).
2) Adsorption of inhibitory substances.
3) Bacterial contamination.
4) Leakage of the enzyme itself.
5) Leakage of enzyme stabilizers.
6) Decomposition of the support material.
7) Disturbance of flow in the column (e.g., channeling).
8) Incorrect operation.
   It is necessary to minimize unfavorable factors to maintain the enzyme activity at a high level.

As an expression of the degree of decay of enzyme activity, i.e., the degree of operational stability, "half-life" is generally used. Half-life is the time required for 50% of the initial enzyme activity to be lost.

The value of the half-life can be calculated from Eqs. 4.3 and 4.4.

$$K_d = \frac{2.303}{\theta} \, \log \frac{E_o}{E_\theta} \tag{4.3}$$

$$t_{1/2} = \frac{0.693}{K_d} \tag{4.4}$$

In both equations, K_d is the enzyme decay constant, θ is the operating time, E_o is the initial enzyme activity, E_θ is the enzyme activity after time θ, and $t_{1/2}$ is the half-life.

Operating temperature markedly influences the stability of enzyme activity. As an enzyme is generally more stable at low temperature, the half-life of enzyme activity increases as the temperature is reduced. However, the enzyme reaction rate also decreases at low temperature. Therefore, the relationship between enzyme reaction rate and operational stability must be considered in the design of enzyme reactors and in process optimization.

For example, the authors[12] investigated the effect of temperature on the stability of fumarase activity of immobilized *Brevibacterium ammoniagenes* cells in continuously flowing substrate solution at 29–45°C. From log plots of remaining activity factor, ϕ, against operating time, θ, the decay of enzyme activity was found to be well-expressed by Eq. 4.3. The results are shown in Fig. 4.6. The enzyme decay constant, K_d, at each temperature was estimated by the least-squares regression method.

The relationship between log K_d and $1/T$ was investigated, and a straight line was obtained, as shown in Fig. 4.7. This indicates that K_d

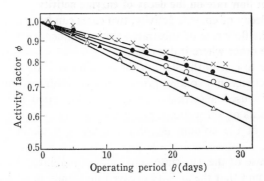

Fig. 4.6 Effect of temperature on the stability of fumarase activity of an immobilized *B. ammoniagenes* cell column.[12]
A solution of 1 M sodium fumarate (pH 7.0) was continuously passed through the column at a flow rate of $SV = 0.71$ at various temperatures. Symbols represent experimental data points. The solid lines show the decay calculated from the K_d values of Fig. 4.7.
Operating temperature: ×—×, 29°C; ●—●, 33°C; ○—○, 37°C; ▲—▲, 41°C; △—△, 45°C.

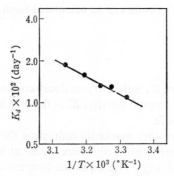

Fig. 4.7   Relationship between temperature and decay constant (K_d).[12]
The K_d values at each temperature were calculated from the results of Fig.
4.6 by the least-squares regression method.

can be expressed by an Arrhenius-type equation (Eq. 4.5),

$$K_d = A_d \cdot \exp\left(-E_d/RT\right) \tag{4.5}$$

where A_d is the frequency factor, R is the gas constant, and E_d is the
deactivation energy. By the least-squares regression method the values of
A_d and E_d were calculated to be 3,788/day and 7,741 cal/mol, respectively.
From Eqs. 4.3 and 4.5, the half-life of fumarase activity of the immobilized *B. ammoniagenes* cells column operated at 37°C was calculated to
be 52.5 days under the conditions employed.

D.   Effect of flow rate on the decay of enzyme activity

In the decay of enzyme activity, two cases can be considered: that is,
one where the flow rate of substrate solution influences the deterioration
rate, and the other where it does not.

In order to clarify the state of the reaction and the decay behaviour
of enzyme activity in a column process using immobilized microbial cells,
the authors[11] used a sectional column, as illustrated in Fig. 4.8. The
column consists of ten sections, and each section has a sampling pit. The
temperature of the column is controlled by recirculating water in the
outer jacket.

In the case of the aspartase activity of immobilized *E. coli* cells, the
authors[11] found that the activity decayed uniformly in all sections of the
column, and the decay of the activity did not depend on the volume-on-
stream of substrate solution, but on the time-on-stream.

On the other hand, in the case of the fumarase activity of immobilized *B. ammoniagenes* cells, the decay of the activity was faster in the
upper sections than in the lower sections.[13] The half-life determined at the
various sections and the remaining activity of the immobilized cells of each

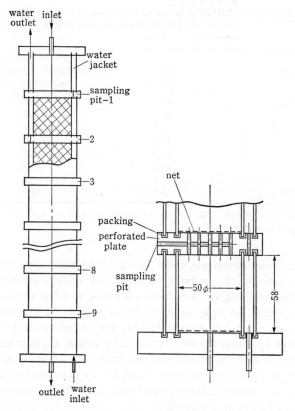

Fig. 4.8 Sectional column.[11)]
Volume, 116 ml/section; length, 5.8 cm/section; diameter, 5.0 cm.

section after 43 days of operation are shown in Table 4.5. The results show that the decay of the enzyme activity does not proceed uniformly throughout the column.

This discrepancy in the two cases is presumably caused by a difference in the sensitivities of aspartase and fumarase to the factors influencing the decay of enzyme activity. The factors causing uneven decay are considered to be as follows.
1) Stability difference in substrate solution and in product solution.
2) Contamination of the enzyme by "poisons" derived from the substrate solution.
3) Leakage of enzyme stabilizer(s) from the immobilized cells.
4) Leakage of enzyme from the immmbilized cells.
5) Irreversible denaturation of enzyme due to its turnover and/or elevation of the heat of reaction.

TABLE 4.5   Half-life of fumarase activity and remaining fumarase
activity of an immobilized *B. ammoniagenes*
cell column

Section No.	Half-life (days)	Remaining activity[†] (formation of L-malic acid μmol/min/g of gel)
1 (top)	49.1	8.68
2	55.9	9.08
3	50.6	9.65
4	57.3	10.48
5	55.0	10.68
6	66.0	10.37
7	61.9	10.68
8 (bottom)	69.3	10.88

† Remaining activity was determined after 43 days of operation. Activity
before operation was 16.34.

To further investigate these factors influencing the decay of the
fumarase activity, the authors[13] carried out various experiments. The
linear velocity of substrate solution was found to affect the decay rate of
enzyme activity. An increase in the linear velocity of substrate solution
results in an increase of mass transfer of the boundary layer between the
immobilized cells and the outer substrate solution. Consequently, higher
linear velocity of the substrate solution may cause an increase in the rate
of contamination of the enzyme with inhibitory substance(s) derived from
the substrate solution or in the rate of leakage of enzyme stabilizers and/or
enzyme itself from the immobilized cells.

Therefore, it is important that the flow rate, especially the linear
velocity, of substrate solution should be taken into consideration in the
optimization of continuous enzyme reactions using immobilized systems.

4.1.5   Factors Affecting Industrial Applications

Based on our experience with industrial applications of immobilized
enzymes and immobilized microbial cells, the following factors are considered to be important: (1) cost of carriers or reagents for the immobilization of enzymes or microbial cells; (2) activity of the immobilized
enzyme and yield from native enzyme or intact cells; (3) stability of the
immobilized enzyme or immobilized cells during operation; (4) regenera-
bility of the deteriorated immobilized enzyme or microbial cells after long
periods of operation.

Besides these, a number of other factors should be considered in industrial applications of immobilized enzymes or immobilized cells, as

TABLE 4.6 Factors to be considered for industrial applications of immobilized enzymes[83]

	Factors	Soluble enzyme (batch system)	Immobilized enzymes Batch system	Immobilized enzymes Column system
Enzyme	cost: high	—	suitable	suitable
	low	suitable	—	—
	reuse	impossible	possible	possible
	stability	low	moderate to high	high
Enzyme reaction	control	difficult	difficult	easy
	rate: high	—	—	suitable
	low	suitable	suitable	—
Product	purity	low	high	high
	yield	low	high	high
Equipment	initial cost	low	moderate	high
	automation	difficult	difficult	easy
	applicability	high	high	moderate
Running costs	labor cost	high	moderate	low
Advantage of scale		low	low	high

shown in Table 4.6. A continuous column system employing an immobilized enzyme is suitable in cases where the cost of the enzyme is high and/or the enzyme reaction rate is high. In the case of a column system, the enzyme reaction can be easily controlled, automation of the process is easy, and the running cost is relatively low. By employing an immobilized enzyme, a product of higher purity can be obtained in higher yield. On the other hands, for the column system, the initial capital cost of equipment is fairly high and the versatility, i.e., ability to use a wide variety of enzyme-substrate combinations, is less than with a batch system; however, advantages on scaling-up can be anticipated.

4.2 CHEMICAL PROCESSES USING IMMOBILIZED SYSTEMS

In this section, applications of immobilized enzymes and immobilized microbial cells for chemical processes are described in the order of classification of enzyme reactions recommended by the enzyme commision of the International Union of Pure and Applied Chemistry and the International Union of Biochemistry in 1972.

Published studies on applications of immobilized enzymes are listed in Table 4.7 in the order of the above recommendations. Examples of chemical processes using immobilized microbial cells so far reported have already been listed in Tables 2.21 through 2.24 and 2.26 in chapter 2.

TABLE 4.7   Application of immobilized enzymes for chemical processes

Catalytic reaction	Immobilized enzyme	Chemical process	Reference
Oxidation-reduction	L-amino acid oxidase	production of D-amino acids	14
	β-tyrosinase	production of L-DOPA	15
		production of L-tyrosine	16, 17
	Δ'-hydrogenase	production of prednisolone	18
	flavoprotein oxidase	N-oxidation of drugs containing amino or hydrazine groups	19
Group transfer	dextransucrase	production of fructose-rich dextran	20
	phosphorylase	polymerization of glucose	21
	polynucleotide phosphorylase	production of polynucleotides	22–24
	carbamate kinase	regeneration of ATP	25
Hydrolysis	ribonuclease	synthesis of tri-nucleotides	24
	α-amylase	production of glucose	26–28
	glucoamylase	production of glucose	26, 28, 29–36
	cellulase	production of glucose	37, 38
	invertase	production of invert sugar	39–43
	leucine aminopeptidase	optical resolution of DL-amino acids	44
	carboxypeptidase	optical resolution of DL-amino acids	45
	papain	hydrolysis of casein	46
	penicillin amidase	production of 6-amino-penicillanic acid	42, 47, 48
		synthesis of penicillins and cephalosporins	49
	aminoacylase	optical resolution of DL-amino acids	2, 50–66
	AMP deaminase	production of 5'-inosinic acid	67
Lyase reactions (asymmetric synthesis)	aspartase	production of L-aspartic acid	10
	tryptophanase	production of L-tryptophan	16, 17
	D-oxynitrilase	productin of D-mandelonitrile	68
Isomerization	glucose isomerase	production of fructose	69–73

4.2.1 Oxidation-reduction Reactions

Regarding the use of immobilized oxidoreductases, the following examples have been reported.

One procedure for the preparation of optically active amino acids involves oxidizing one antipode of DL-amino acid with an amino acid oxidase. In order to carry out this reaction continuously using immobilized enzyme, L-amino acid oxidase immobilized on porous glass was packed into a column, and DL-amino acid was passed through the column.

The L-amino acid was converted to α-keto acid, and the unsusceptible D-amino acid was isolated from the effluent.[14] Further, as shown in Eq. 4.6, the immobilized L-amino acid oxidase was also used to prepare L-tyrosine by transamination between DL-phenylalanine and p-hydroxyphenylpyruvic acid under anaerobic conditions.[74]

$$
\underset{\text{DL -phenylalanine}}{\text{C}_6\text{H}_5\text{-CH}_2\text{CHCOOH (NH}_2\text{)}} + \underset{p\text{-hydroxyphenylpyruvic acid}}{\text{HO-C}_6\text{H}_4\text{-CH}_2\text{COCOOH}} \xrightarrow[\text{(under anaerobic conditions)}]{\text{L- amino acid oxidase}}
$$

$$
\underset{\text{phenylpyruvic acid}}{\text{C}_6\text{H}_5\text{-CH}_2\text{COCOOH}} + \underset{\text{L - tyrosine}}{\text{HO-C}_6\text{H}_4\text{-CH}_2\text{CHCOOH (NH}_2\text{)}} + \underset{\text{D - phenylalanine}}{\text{C}_6\text{H}_5\text{-CH}_2\text{CHCOOH (NH}_2\text{)}} \qquad (4.6)
$$

Although the reaction proceeded with soluble enzyme, the apparent conversion rate by the immobilized enzyme in a closed column system was reported to be about 10 times higher than that by the soluble enzyme.[74]

In the case of 3,4-dihydroxy-L-phenylalanine (L-DOPA), which is used for treating Parkinson's disease, its preparation from L-tyrosine using immobilized tyrosinase prepared by covalent binding to the triazine derivative of DEAE-cellulose was studied. Namely, the immobilized enzyme was packed into a column and L-tyrosine solution containing ascorbic acid was passed through the column to give L-DOPA.[15] In this case, ascorbic acid was added to reduce the formation of quinone by further oxidation of L-DOPA. As the mushroom tyrosinase used for this reaction is strongly inhibited by the product, L-DOPA, studies are required to find other sources of enzyme which does not suffer from product inhibition.

An attempt to use immobilized tyrosinase more directly for therapy of Parkinson's desease by conversion of L-tyrosine in serum to L-DOPA by using the immobilized enzyme in an extracorporeal shunt system was also made, as described later (section 4.4).[15]

Hepatic flavoprotein oxidase has also been immobilized by covalent binding to CNBr-activated Sepharose or to partially hydrolyzed nylon tube, as well as to aminoalkylated glass beads with glutaraldehyde. These immobilized oxidases were used for the preparation of N-oxidized metabolites of clinically useful drugs such as ethylmorphine, chloropromazine, prochlorperazine, etc.[19] These N-oxidized metabolites are difficult to prepare by direct chemical oxidation, and only small amounts of them can be isolated from urine after in vivo administration of these drugs.

In addition, for the oxidation, reduction and hydroxylation of steroid hormones, methods using highly specific enzymes or microbial cells have been widely studied, and some of them are said to be in use industrially.

In these conversion reactions of steroids, Mosbach and Larsson[18] used both immobilized lichen cells and immobilized enzyme for the conversion of 11-deoxycortisone (1) (Reichstein compound S) to cortisol (2) and further to prednisolone (3), as shown in Eq. 4.7. Lichen cells belonging to *Curvularia lunata* were used as an enzyme source of 11 β-hydroxylase

$$(4.7)$$

catalyzing the first reaction, after immobilization using polyacrylamide gel. This immobilized *C. lunata* was suspended in physiological saline solution containing Tween 80, and 11-deoxycortisone (1) dissolved in dimethyl sulfoxide was added. The reaction was carried out with shaking in the dark at 28°C to give cortisol (2). It is interesting that immobilized *C. lunata* which had deteriorated on storage was claimed to be regenerated by shaking with corn steep liquor, sugar, sodium chloride, Tween 80 and cortisol (**2**).

Further, the cortisol obtained was converted to prednisolone (**3**) by passage through a column packed with immobilized Δ^1-dehydrogenase. This enzyme was extracted from *Corynebacterium simplex*, partially purified, and immobilized by the polyacrylamide gel method. This technique is interesting as an application of a multienzyme system in which immobilized enzyme and immobilized cells are well combined.

4.2.2 Transfer Reactions

In this section, applications of immobilized enzymes mainly for the synthesis of macromolecular substances such as polysaccharides and nucleic acids are described.

As an example of the synthesis of polysaccharide, the formation of fructose-rich dextran from sucrose solution of high concentration by using dextransucrase bound ionically to DEAE-Sephadex was reported.[20] The dextran obtained could interact with concanavalin A and exhibited a higher glycogen value and fructose content than that synthesized by the soluble enzyme.

Further, by using phosphorylase bound covalently to porous glass with glutaraldehyde, a polysaccharide with a molecular weight of 20,000–70,000 was prepared according to Eq. 4.8.[21]

$$x \text{ D-glucose-1-phosphate} + (\text{D-glucose})_n$$
$$\xrightarrow{\text{phosphorylase}} (\text{D-glucose})_{n+x} + x \text{ inorganic phosphate} \tag{4.8}$$

As examples of polynucleotide synthesis, papers have appeared concerning immobilized polynucleotide phosphorylase.[22,23] Namely, by using enzyme immobilized by binding to CNBr-activated cellulose, poly A, U, I and C were prepared from the corresponding diphosphates, and the biological activities and physical characteristics of these homopolymers were examined.[22,23] Further, the enzyme immobilized by binding to CNBr-activated Sepharose was packed into a column and a solution of ADP, CDP, or 5-Cl-CDP containing $MgCl_2$ and EDTA was passed through the column. The effluents were dialyzed to give the corresponding homopolymers, poly A, poly C or poly 5-Cl-C, of high molecular weight (sedimentation constant, 6–125) in high yield.[23]

Further, as shown in Eq. 4.9, commercially available immobilized ribonucleases prepared on cellulose azide and maleic copolymer were used

$$\text{cyclic Up} + \begin{cases} \text{A–A} \longrightarrow \text{U–A–A} \\ \text{A–G} \longrightarrow \text{U–A–G} \\ \text{G–A} \longrightarrow \text{U–G–A} \end{cases} \tag{4.9}$$

for the synthesis of trinucleotides corresponding to termination codons in protein biosynthesis.[24]

Immobilized carbamate kinase prepared by covalent binding to porous glass with glutaraldehyde was packed into a column, and regeneration of ATP was carried out on the column according to Eqs. 4.10 and 4.11. This immobilized enzyme was stable and retained 74% of the initial activity after operation for 14 days at room temperature.[25]

$$KH_2PO_4 + KOCN + H_2O \xrightarrow{\text{nonenzymic}} [NH_2COOPO_3H^-]K^+ + KOH \qquad (4.10)$$

$$[NH_2COOPO_3H^-]K^+ + ADP \xrightarrow{\text{carbamate kinase}} ATP + [NH_2COO^-]K^+ \qquad (4.11)$$

4.2.3   Hydrolytic Reactions

Many kinds of hydrolases and microbial cells having their enzyme activities have been immobilized, and they have been used as solid catalysts for chemical processes. The reasons for this extensive use are considered to be as follows: 1) there are many different kinds of hydrolases; 2) hydrolases are generally stable; 3) hydrolases do not require coenzymes; 4) hydrolases become more stable on immobilization and can be used for a long time.

The authors[51] have been employing immobilized aminoacylase for the industrial production of L-amino acids. Other hydrolytic enzymes are expected to be used for industrial purposes in the near future.

In this section, examples of applications of immobilized hydrolases and immobilized microbial cells are discussed from the viewpoint of the enzymes or reaction products.

A.   Production of L-amino acids
a.   Production of L-Amino Acids by Immobilized Aminoacylase

Utilization of L-amino acids in medicine, food and animal feed has been developing rapidly in recent years, and the economical production of optically active amino acids has been investigated extensively.

At present, fermentative and chemical synthetic methods are employed for the industrial production of L-amino acids instead of conventional isolation from protein hydrolysates. However, chemically synthesized amino acids are optically inactive racemic mixtures of L- and D-isomers. The L-form is the physiologically active natural form, and is therefore the required form for medicine and food. To obtain L-amino acid from chemically synthesized DL-form, optical resolution is necessary.

Generally, optical resolution of racemic amino acids can be carried out by physicochemical, chemical, enzymic, and biological methods. Among these methods, an enzymic method, developed by the authors, using mold aminoacylase is one of the most advantageous procedures, yielding optically pure L-amino acids. The reaction catalyzed by the enzyme is shown in Eq. 4.12.

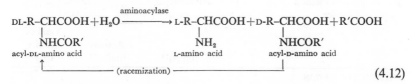

Chemically synthesized acyl-DL-amino acid is asymmetrically hydrolyzed by aminoacylase to give L-amino acid and the unhydrolyzed acyl-D-amino acid. After concentration, both materials are easily separated on the basis of the difference in their solubilities. Acyl-D-amino acid is racemized, and reused for the resolution procedure.

The authors[75-77] studied enzymes catalyzing the reaction of Eq. 4.12 and found that aminoacylase produced by *Aspergillus oryzae* has high activity and broad substrate specificity, that is, it can asymmetrically hydrolyze many kinds of acyl–DL-amino acids.

From 1953 to 1969, the authors employed this mold aminoacylase for the industrial production of several L-amino acids. This method was one of the most advantageous procedures for the industrial production of L-amino acids. However, the enzyme reaction was carried out batchwise by incubating a mixture containing the substrate and soluble enzyme. Thus, the procedure had some disadvantages for industrial purposes. For instance, in order to isolate L-amino acid from the enzyme reaction mixture, it was necessary to remove enzyme protein by pH and/or heat treatment. Thus, even if enzyme activity remained in the reaction mixture, the enzyme had to be discarded because there was no suitable procedure for isolating the active enzyme from the mixture. In addition, a complicated purification procedure was necessary for the removal of proteins and coloring materials contaminating in the crude enzyme preparations usually employed for industrial purposes. As a result, the yield of L-amino acids was reduced. Also, a considerable amount of labor was necessary for batch operation.

To overcome these disadvantages and to improve this enzymic method, the authors extensively studied the continuous optical resolution of DL-amino acids using a column packed with immobilized amino-acylase. The following immobilization methods were extensively tested: [2,8,51~53,55,56,78~82] physical adsorption,[51] ionic binding,[51] covalent binding,[80] cross-linking,[79] and entrapping.[81] These immobilized amino-acylases are summarized in Table 4.8. It was found that 3 kinds of immobilized preparations obtained by ionic binding to DEAE-Sephadex, covalent binding to iodoactylcellulose and entrapping in a polyacrylamide gel

TABLE 4.8   Relationship between immobilization and yield
of activity of aminoacylase[83]

Immobilization method and carrier	Aminoacylase used	Immobilized aminoacylase	
	(units)[†1]	Activity (units)[†1]	Yield of activity (%)
PHYSICAL ADSORPTION			
Acid aluminum oxide	1210	13	1.0
Neutral aluminum oxide	1210	10	0.8
IONIC BINDING			
DEAE-cellulose	1210	668	55.2
ECTEOLA-cellulose	1210	293	24.2
TEAE-cellulose	1210	623	51.5
DEAE-Sephadex A-25	1210	713	58.9
DEAE-Sephadex A-50	1210	680	56.2
COVALENT BINDING			
Diazotized PAB cellulose	1210	64	5.3
Diazotized Enzacryl AA	1210	44	3.6
Diazotized arylamino glass	1210	525	43.4
CNBr-activated cellulose	1210	12	1.0
CNBr-activated Sephadex	1210	15	1.2
Chloroacetyl-cellulose	1210	137	11.3
Bromoacetyl-cellulose	1210	339	28.0
Iodoacetyl-cellulose	1210	472	39.0
CARRIER CROSS-LINKING			
Glutaraldehyde	1440	8	0.6
Hexamethylenediisocyanate	1440	23	1.6
CROSS-LINKING			
Glutaraldehyde	1440	211	14.7
Toluene diisocyanate	1440	18	1.3
LATTICE ENTRAPPING			
Polyacrylamide gel	1000	526	52.6
HPMCP-DEAE[†2]	1000	190	19.0
MICROCAPSULE ENTRAPPING			
Nylon	1000	360	36.0
Polyurea	1000	150	15.0
Ethylcellulose	1000	104	10.4

[†1] One enzyme unit is defined as the amount of enzyme which liberated 1 μmol of L-methionine per hour at 37°C.

[†2] Diethylaminoethyl derivative of hydroxypropiomethyl-cellulose phthalate.

lattice were relatively favorable as regards yield, activity and stability. However, for the industrial operation of a continuous enzyme reaction using an immobilized enzyme, it was necessary to satisfy many conditions. Thus, the characteristics of these three immobilized aminoacylases were compared, and aminoacylase bound to DEAE-Sephadex was chosen as the most advantageous enzyme preparation for the industrial production of L-amino acids, because 1) preparation is easy, 2) the activity is high, 3) the stability is high and 4) regeneration of deteriorated immobilized enzyme is possible.

In order to carry out a continuous enzyme reaction using the immobilized enzyme, it was necessary to design a suitable enzyme reactor as described in section 4.1.

In this section, some typical experimental data on the immobilized aminoacylase are described.

As one of the factors to be considered in the design of an enzyme reactor, it is necessary to investigate in advance the relationship between the flow rate of substrate solution passing through the enzyme column and the extent of the reaction. The authors[55] investigated the relationship between the flow rate of a solution of acetyl–DL-methionine or acetyl–DL-phenylalanine and the extent of the reaction, and the results are shown in Fig. 4.9. That is, an aqueous solution of acetyl–DL-amino acid (0.2 M, pH 7.0–7.5, containing 5×10^{-4} M Co^{2+}) was passed through a DEAE–Sephadex–aminoacylase column at various flow rates at 50°C. In this figure, flow rate is expressed in terms of space velocity (SV). The space velocity is the volume of liquid passing through a given volume of immobilized enzyme in 1 h divided by the latter volume. If the space velocity

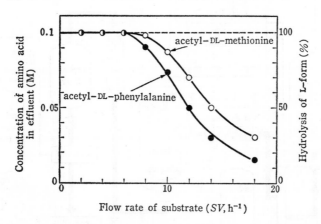

Flow rate of substrate (SV, h^{-1})

Fig. 4.9 Continuous hydrolysis of acetyl-DL-amino acids by an immobilized aminoacylase column—the relationship between flow rate of substrate and extent of hydrolysis.[55]
Aminoacylase was immobilized by ionic binding to DEAE-Sephadex.
Concentration of substrate, 0.2 M (containing 5×10^{-4} M Co^{2+}); reaction temperature, 50°C.

is reduced, that is, the flow rate is slowed down, the reaction proceeds more effectively; the relationship between the flow rate and the extent of the reaction is sigmoidal. As shown in Fig. 4.9, when the concentration of L-amino acids in the effluent reached 0.1 M, the reaction stopped. This result indicated that the L-form of 0.2 M acetyl–DL-amino acid was completely hydrolyzed and its D-form was not hydrolyzed, that is, asymmetric hydrolysis occurred.

The pressure drop of the enzyme column is also important factor in the design of an enzyme reactor, as described in section 4.1. Thus, the relationship between the flow rate of substrate solution and the pressure drop of the column was investigated using columns of various lengths. The pressure drop of the DEAE-Sephadex-aminoacylase column was found to be proportional to the flow rate of substrate solution and the column length.

In addition, by considering some other properties of the immobilized enzyme, such as optimum pH, the effect of temperature on the reaction rate and stability, and so on, the authors[79] designed an enzyme reactor system for continuous production of L-amino acids as shown in Fig. 4.10. In this system, the flow rate and pH of substrate solution, and operating temperature can be automatically controlled and recorded.

Since 1969, continuous optical resolution of acetyl–DL-amino acids has been carried out using this sytem and several kinds of optically active

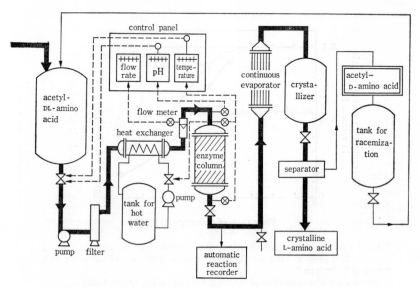

Fig. 4.10   Flow diagram for the continuous production of L-amino acids by immobilized aminoacylase.[79]

amino acids such as methionine, phenylalanine, valine and others have been industrially produced at Tanabe Seiyaku Co. Ltd., Japan.

Several examples of the production of L-amino acids are summarized in Table 4.9, which shows the space velocity and the theoretical yield for each amino acid produced in 1000-liter aminoacylase column. That was the first industrial application of immobilized enzymes.[83]

The enzyme column and the control panel for continuous operation installed in the plant of Tanabe Seiyaku Co. Ltd. are shown in Photos 4.1 and 4.2, respectively.

TABLE 4.9 Production of L-amino acids on a 1000 liter DEAE-Sephadex aminoacylase column[78]

L-Amino acid	Space velocity	Yield (theory) of L-amino acids per	
		24 h (kg)	20 days (kg)
L-Alanine	1.0	214	6,420
L-Methionine	2.0	715	21,450
L-Phenylalanine	1.5	594	17,820
L-Tryptophan	0.9	441	13,230
L-Valine	1.8	505	15,150

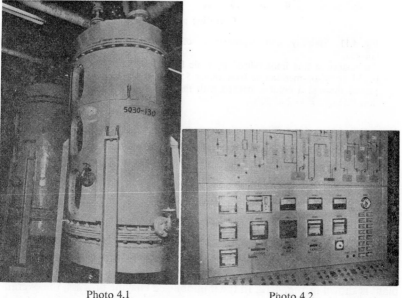

Photo 4.1   Photo 4.2

Photo 4.1 Immobilized aminoacylase column for continuous enzyme reaction (Tanabe Seiyaku Co. Ltd.).

Photo 4.2 Control panel for continuous enzyme reaction (Tanabe Seiyaku Co. Ltd.).

This immobilized aminoacylase was very stable, as shown in Fig. 4.11 and maintained 60~70% of the initial activity after continuous operation for 30 days at 50°C. The enzyme column can be completely regenerated after prolonged operation by the addition of an amount of aminoacylase corresponding to the lost activity. The carrier, DEAE–Sephadex, is also very stable, and has been used for over 5 years in our column process without any change in adsorption capacity for the enzyme, shape or pressure

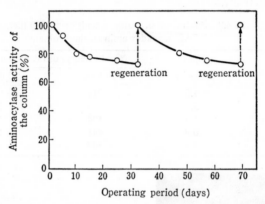

Operating period (days)

Fig. 4.11 Stability and regeneration of an immobilized aminoacylase column.[79]
Aminoacylase was immobilized by ionic binding to DEAE-Sephadex, and 0.2 M acetyl-DL-methionine (containing 5×10^{-4} M Co^{2+}) was continuously passed through a column packed with the immobilized aminoacylase at a flow rate of $SV = 2$ at 50°C.

drop. Continuous enzyme reaction using such a stable immobilized enzyme not only gives high productivity per enzyme unit, but also high product yield, because contamination of the effluent by impurities such as proteins and coloring substances does not occur, so the product can be obtained by simple purification and the amount of substrate required is reduced. Further, automatic operation reduces labor costs and a significant reduction of production cost can be expected in comparison with the conventional batch process using soluble enzyme. A comparison of the production costs of L-amino acids by the conventional batch process using soluble enzyme and by the continuous process using immobilized enzyme in our plant is shown in Fig. 4.12. In the immobilized enzyme process, the overall production cost is more than 40% lower than that of the conventional batch process using soluble enzyme. Savings of enzyme and labor costs are the main contributors, as well as the increase of product yield due to easy isolation of L-amino acids from the reaction mixture. Although

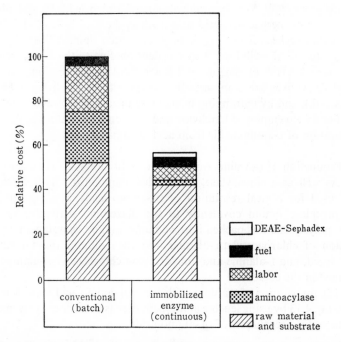

Fig. 4.12 Comparison of production costs of L-amino acids by batch and continuous processes.[79]

DEAE-Sephadex is a relatively expensive carrier, production costs are not greatly affected by the cost of the carrier, because it can be used for a very long period, as mentioned above.

This procedure can be applied for the production of a number of amino acids and is especially advantageous in the case of amino acids which are difficult to produce by fermentation. Further, the same method can be used for the production of D-amino acids by chemical hydrolysis of acetyl–D-amino acids obtained after the enzyme reaction. In practice, several D-amino acids have been industrially produced by this procedure.

Besides the above immobilization by ionic binding, mold aminoacylase from *Aspergillus* sp. was also immobilized by covalent binding to alkylaminosilanized porous glass with glutaraldehyde or to the diazonium derivative of arylaminosilanized porous glass, and these immobilized aminoacylases were used for continuous preparation of L-amino acids from acetyl–DL-amino acids.[58]

On the other hand, bacterial and animal aminoacylases have been also immobilized as follows.

Bacterial aminoacylase has been covalently bound to diazotized polyaminostyrene,[59~61] N-carboxyanhydrides of amino acids,[62] and nitrated copolymer of methacrylic acid and methacrylic acid-3-fluoroanilide.[63,64] The aminoacylase from pig kidney has been immobilized by ionic binding to DEAE-cellulose,[54] by covalent binding to the azide derivative of Enzacryl AH[50] or diazotized Enzacryl AA[65] (see section 2.1.1; Enzacryl derivatives are commercially available from Koch-Light Laboratories Ltd.), and by entrapping in cellulose nitrate.[66] However, these preparations were employed batchwise and only on a laboratory scale for the preparation of L-amino acids from acyl–DL-amino acids.

b.   Production of L-Amino Acids by Other Immobilized Hydrolases

As well as aminoacylase, some other immobilized hydrolases have been used for optical resolution of DL-amino acids. For example, carboxypeptidase bound covalently to the diazonium derivative of polyaminopolystyrene was packed into a column, and on charging an aqueous solution of chloroacetyl–DL-alanine into the column, the L-form was hydrolyzed, and L-alanine and unhydrolyzed chloroacetyl–D-alanine were obtained in the effluent.[45]

The preparation of L-amino acids from DL-amino acid amides according to Eq. 4.13 by using leucine aminopeptidase is another enzymic method for the optical resolution of DL-amino acids.

$$\begin{array}{ccc} & \text{leucine aminopeptidase} & \\ \text{DL-R–CHCONH}_2 + \text{H}_2\text{O} \xrightarrow{\hspace{2cm}} & \text{L-R–CHCOOH} + \text{D-R–CHCONH}_2 \\ | & | & | \\ \text{NH}_2 & \text{NH}_2 & \text{NH}_2 \\ \text{DL-amino acid amid} & \text{L-amino acid} & \text{D-amino acid amide} \end{array}$$

$$(4.13)$$

Leucine aminopeptidase immobilized by covalent binding to CNBr-activated Sepharose was packed into a column and continuous enzyme reaction was carried out to give L-alanine, L-leucine, L-phenylalanine, and so on.[44]

c.   Production of L-Citrulline by Immobilized *Pseudomonas putida*

L-Citrulline is used in medicines, and is industrially produced from L-arginine by the action of L-arginine deiminase, as shown in Eq. 4.14.

$$\begin{array}{ccc} \text{NH} & \begin{array}{c}\text{L-arginine}\\\text{deiminase}\end{array} & \text{O} \\ || & & || \\ \text{H}_2\text{NCNHCH}_2\text{CH}_2\text{CH}_2\text{CHCOOH} \xrightarrow{\hspace{1.5cm}} & \text{H}_2\text{NCNHCH}_2\text{CH}_2\text{CH}_2\text{CHCOOH} + \text{NH}_3 \\ | & | \\ \text{NH}_2 & \text{NH}_2 \\ \text{L-arginine} & \text{L-citrulline} \end{array}$$

$$(4.14)$$

Many kinds of microorganisms having this enzyme activity have been reported, and in many cases, these microorganisms have enzyme systems further metabolizing the resulting L-citrulline. The authors[84] found that *Pseudomonas putida* has a high L-arginine deiminase activity and has no activity degrading L-citrulline. The microorganisms has been industrially used for the production of L-citrulline at Tanabe Seiyaku Co. Ltd.

Thus, the authors[85] investigated the conditions for continuous production of L-citrulline by this immobilized microorganism. When the immobilization of *P. putida* cells was carried out by the polyacrylamide gel method, the most active immobilized cells were obtained. Some of the enzymatic properties of the immobilized cells were investigated and compared with those of the intact cells. It was found that there was no difference between the pH-activity curves of the intact and immobilized cells, but the heat stability and optimum temperature of the enzyme activity were increased by immobilization of the cells. In the intact cells of *P. putida*, the reaction did not occur unless a cationic surfactant such as cetyltrimethylammonium bromide was added to the reaction mixture for L-citrulline formation, because the permeability of the substrate or product through the cell membrane was low. However, in the case of the immobilized cells, the reaction proceeded without addition of surfactants. This phenomenon indicates that the membrane of intact cells is a barrier to L-arginine or L-citrulline and that immobilization of the cells possibly removes this barrier.

When an L-arginine solution (pH 6.0) was passed through a column packed with the immobilized cells, the reaction proceeded in the column by the action of L-arginine deiminase according to Eq. 4.14. After com-

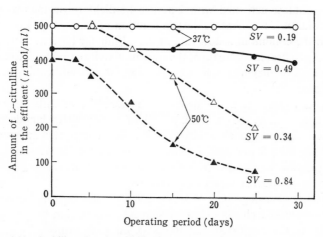

Fig. 4.13   Stability of an immobilized *P. putida* column.[85]
A solution of 0.5 M L-arginine (pH 6.0) was continuously passed through the immobilized cell column at the indicated flow rate at 37°C or 50°C.

pletion of the reaction, the effluent of the column contained only equimolar ammonium chloride and L-citrulline. L-Citrulline is easily obtained in good yield by concentration of the effluent. Fig. 4.13 shows the results of studies on the stabilities of immobilized cell columns. When the reaction was carried out at 50°C, a rapid decrease of the activity was observed. However, when the reaction was carried out continuously at 37°C for about 30 days, no significant decrease of the activity was observed, and the half-life was calculated to be about 140 days.

B. Hydrolysis of starch (production of glucose)

At present, glucose is mainly produced by the so-called enzymic method, in which starch is hydrolyzed by the action of α- or β-amylase and glucoamylase. In order to carry out this process continuously, procedures using immobilized enzymes or soluble enzymes in enzyme reactors having ultrafiltration membranes have been extensively investigated in the U.S.A., Japan, England and others.[26,29~35,42,86~88]

For this purpose, α-amylase and glucoamylase have been immobilized by various methods, as described in chapter 2. For instance, glucoamylase was immobilized by covalent binding to CM–cellulose azide[29,36] or DEAE–cellulose activated with 2-amino-4,6-dichloro-s-triazine,[31,36] by ionic binding to DEAE–cellulose[32,33] and by physical adsorption to activated carbon.[89] The immobilized enzymes were packed into a column or continuous reaction vessel, and suitable conditions for the hydrolysis of starch were investigated.

For example, glucoamylase immobilized by covalent binding to DEAE–cellulose activated with 2-amino-4,6-dichlorotriazine was packed into a column (5 × 13 cm). A dextrin solution of 53.7% (w/w) was passed through the column at a flow rate of 2.8 ml/min for 100 h at 55°C to give 3.8 kg of glucose; no decrease of the enzyme activity could be detected.[30,36] The kinetics of enzyme reactors for continuous saccharification were also investigated.[31]

Besides these studies, glucoamylase alone[35] and with α-amylase[26] was immobilized by the entrapping method using polyacrylamide gel, and conditions for continuous saccharification were investigated in detail using these immobilized preparations. In the case of the latter, when a corn starch solution was passed through a column packed with both immobilized enzymes, glucose syrup was obtained.

Further, Corning Glass Works studied the design of industrial plant using immobilized glucoamylase prepared by adsorption on porous SiO_2 ceramics. It was calculated that 10,000,000 lb (2,222 tons) of glucose per year could be produced by using a 2.5 ft³ column (6 in × 8 ft). A pilot plant with multistage columns of 1 ft³ was operated at Iowa State University, and 1000 lb (222 kg) of glucose per day was produced. During 70 days of operation, no detectable change of the enzyme activity occurred.[90]

However, in the process using immobilized glucoamylase, the pressure drop of the enzyme column increased because of the high viscosity of liquefied starch, channeling was liable to occur in the column, and also it appeared to be difficult for the immobilized enzyme to interact with high-molecular substrate. Because of these technical difficulties, this process has not been industrialized so far.

To overcome these problems, continuous saccharification of starch using a stirred reaction vessel has also been studied. For instance, the flow diagram shown in Fig. 4.14 was designed.[33] Namely, glucoamylase from *Aspergillus awamori* was immobilized by ionic binding to DEAE–cellulose.

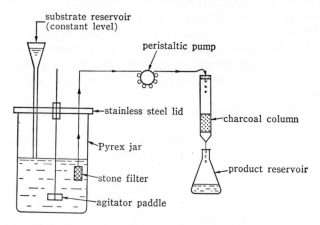

Fig. 4.14 Flow diagram for continuous saccharification of starch using a stirred reactor.[33]
Glucoamylase was immobilized on DEAE-cellulose by the ionic binding method.

The mixture of immobilized enzyme and an aqueous solution of starch partially hydrolyzed with starch-liquifying enzyme was incubated in a stirred reactor, and low-molecular-weight glucose produced by the reaction was continuously withdrawn through a porous stone filter. This immobilized enzyme was very stable and no decrease of activity was observed after operation for 3–4 weeks at 55°C.

Further, reactors using recently developed ultrafiltration membranes have been studied for continuous saccharification.[28] For example, as shown in Fig. 4.15, starch solution (B) kept at constant temperature was transferred to a reaction vessel (C) containing enzyme solution by pressure from a nitrogen cylinder (A), and the enzyme reaction was carried out

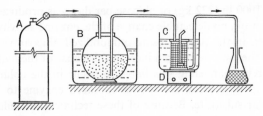

Fig. 4.15   System for continuous hydrolysis of starch using an ultrafiltration membrane.[28]
A, Cylinder of compressed gas; B, reservoir of starch solution; C, reaction vessel; D, magnetic stirrer.

with stirring using a magnetic stirrer (D). Since the membrane on the bottom of the reaction vessel retains macromolecular substances such as the substrate (starch in this case) or enzyme (α-amylase and glucoamylase), only small-molecular hydrolyzed products can pass through.

Further, to hydrolyze corn starch by means of α-amylase, the flow diagram using an ultrafiltration membrane shown in Fig. 4.16 was designed and chemical engineering studies for continuous saccharification were carried out.[27]

The application of ultrafiltration membranes is limited to cases where

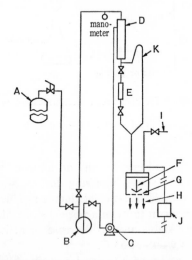

Fig. 4.16   System for continuous enzyme reaction using an ultrafiltration membrane with liquid level control.[27]
A, Nitrogen gas cylinder; B, substrate reservoir; C, pump; D, inverted surge tank; E, sample return port; F, magnetic stirrer; G, ultrafiltration membrane; H, ultrafiltration product; I, liquid level probe; J, liquid level control; K, siphon line.

the substrates are macromolecular, the products are low-molecular and the enzymes are stable. However, this method has some advantages: the process for immobilization of the enzyme can be omitted, and steric hindrance is less in hydrolytic reactions of macromolecular substrates than in the case of immobilized enzymes prepared by the carrier-binding method. Accordingly, the use of stirred reactors having ultrafiltration membranes appears to have a promising future for industrial continuous enzyme reactions.

As already mentioned, since enzymes in the soluble state are relatively unstable compared to immobilized enzymes, it may be advantageous to carry out reactions in a vessel having an ultrafiltration membrane after stabilizing the enzymes by chemical modification. For instance, it was reported that when an enzyme is covalently bound to a water-soluble macromolecular substance such as dextran, its stability increases even in the soluble state and the efficiency is increased.[91]

C. Hydrolysis of cellulose (production of glucose)

Production of glucose from cellulose by enzymic methods has been investigated, and a procedure using an enzyme reactor having an ultra-filtration membrane has been reported for this purpose.[37,38] Cellulose was heated at 200°C in a running porcelain pot mill for 25 min, ground to fine particles with a diameter of 50–150 μ, and then suspended in water. The cellulose suspension was reacted with cellulase from *Trichoderma viride* in an ultrafiltration membrane reactor, and the resulting glucose syrup was continuously removed from the reactor through the membrane.[37] The procedure was carried out on a laboratory scale.

It is also known that cellulase is strongly adsorbed on cellulose and becomes immobilized. It can then hydrolyze cellulose at pH 4–5 and at 25–50°C. With this immobilized enzyme, hydrolysis of cellulose has also been carried out in a column or continuously stirred tank reactor. When cellulose was continuously fed into a reactor containing cellulase, the cellulose was hydrolyzed to give glucose. In this case, even if the glucose is continuously removed from the reactor, the enzyme does not leak out, and continuous reaction can be carried out for a long period because the cellulase is so strongly bound to cellulose.[38]

It is expected that saccharification of cellulose by immobilized cellulase may be carried out more advantageous by using soluble or colloidal cellulose.

D. Production of invert sugar

For the continuous production of invert sugar from sucrose, im-mobilization of invertase has been investigated. For example, invertase from yeast was immobilized by the diazo method using porous glass,[39] by the ionic binding method using DEAE–cellulose,[40] by entrapping methods using polyacrylamide gel[41] and cellulose acetate,[42] by aggregation with tannic acid,[43] and others. Using these immobilized enzymes, continuous hydrolysis of sucrose has been carried out by the column method.

The immobilized enzymes were relatively stable. For instance, in the case of the preparation obtained by covalent binding on porous glass, no significant decrease of activity was observed after continuous operation for 28 days, and its half-life was calculated to be 42.5 days.[39]

In the case of the immobilized enzyme prepared using polyacrylamide gel, practically no decrease of activity was observed after continuous reaction for a week.[41] In this case, substrate solution was charged into the column by upward flow, because when it was charged by downward flow, the packed bed of the column was gradually compressed.[41]

Further, continuous hydrolysis of sucrose by the ultrafiltration membrane method was investigated, and the same continuous reaction system shown in Fig. 4.15 was designed.[92]

However, the production procedure for invert sugar by using immobilized invertase cannot compete with the isomerization method (glucose to fructose) using glucose isomerase as described in section 4.1.5, and it is not used practically at present.

E.   Production of 6-aminopenicillanic acid

6-Aminopenicillanic acid (6-APA) is used as an important intermediate for synthetic penicillin, and has been produced industrially from penicillin by the action of penicillin amidase (penicillin acylase) by a batch method using microbial cells or extracted enzyme as shown in Eq. 4.15.

$$RCONHCH-\underset{\substack{|\\ CO-N}}{\overset{S}{CH}}\overset{\substack{CH_3\\ C}}{\underset{CH_3}{}} + H_2O \xrightarrow{\substack{penicillin\\ amidase}} H_2NCH-\underset{\substack{|\\ CO-N}}{\overset{S}{CH}}\overset{\substack{CH_3\\ C}}{\underset{CH_3}{}} + RCOOH$$

penicillin   6–APA   (4.15)

Recently, many studies have been carried out on the continuous production of 6-APA using a column packed with immobilized penicillin amidase.[42,47,48] For example, penicillin amidase extracted from *Escherichia coli* was immobilized by covalent binding to DEAE–cellulose activated with 2,4-dichloro-6-carboxymethylamino-*s*-triazine.[47] The enzyme is relatively unstable in the soluble state, but its stability is increased by immobilization. The immobilized enzyme can be used for continuous operation without significant loss of activity at 37°C for 11 weeks.

An advantage of this method is that proteins and other impurities causing allergic reaction are not included in the 6-APA obtained. Therefore, the 6-APA can be used advantageously as an intermediate for synthetic penicillins.

A penicillin amidase adsorbed onto bentonite is also used for the industrial production of 6-APA at Squibb, U.S.A.[48] It was reported that the enzyme reaction was inhibited by the hydrolytic products of penicillin G, i.e., 6-APA and phenylacetic acid. However, if continuous enzyme reaction was carried out using a column packed with the immobilized enzyme, the reaction could proceed, because the reactants were removed from the reaction system. This is a major advantage of using the immobilized enzyme for industrial production of 6-APA.

As described previously, if 6-APA was produced by the conventional enzymic method instead of the immobilized enzyme method, trace amounts of proteins causing antigenicity or immunity were sometimes included. To remove such trace amounts of contaminating proteins, immobilized proteases have been used. For instance, it was reported that 6-APA having little or no antigenicity can be produced by hydrolysis of the proteins with an immobilized pronase prepared by covalent binding to bromoacetyl-cellulose or CM–Sephadex.[93]

At present, industrial production of 6-APA by immobilized penicillin amidase is carried out by several companies, such as Astra Lakemedel, Sweden, Beecham, England, and Toyo Jozo Co., Ltd., Japan, as well as Squibb, U.S.A.

More recently, utilizing the reverse reaction of penicillin amidase, synthesis of penicillin and cephalosporin derivatives by immobilized penicillin amidase has been studied,[49] and further developments are expected.

In addition, to find a more advantageous method for the industrial production of 6-APA from penicillins, the authors[94] investigated the immobilization of *E. coli* cells having penicillin amidase activity by the polyacrylamide gel method.

The enzymic properties of immobilized cells were investigated and compared with those of intact cells. The results showed that the heat stability of the enzyme was increased by immobilization.

Using a column packed with the immobilized *E. coli* cells, continuous production of 6-APA from penicillin G was investigated. These microbial cells contain penicillinase, which decomposes both penicillin and 6-APA, and selective inactivation of penicillinase activity is very difficult. However, penicillinase activity is much lower than penicillin amidase activity. Therefore, optimum conditions for the continuous production of 6-APA without destroying the penicillinase activity were selected. It was found that when 0.05 M penicillin G solution (pH 8.5) was passed through the immobilized *E. coli* cell column at 40°C at a space velocity of 0.24, 6-APA was efficiently produced. From the effluent 6-APA was obtained in about 80% yield. Therefore, this technique is considered to be useful for the industrial production of 6-APA, as well as the continuous method using immobilized penicillin amidase.

F.   Miscellaneous hydrolytic reactions

In addition to the enzymes described above, other hydrolases have been used for continuous enzyme reactions after their immobilization. For example, AMP deaminase from *Aspergillus melleus* immobilized by ionic binding to DEAE–cellulose was packed into a column, and an aqueous solution of 5'-adenylic acid was passed through the column, yielding 5'-inosinic acid efficiently.[67] The activity of the enzyme column showed no detectable decrease during 3 days of operation and after operation for a further 2 days its activity was about 70% of the initial value.

Further, papain immobilized by covalent binding to glass beads treated with zirconium oxide has been used for continuous hydrolysis of casein by the column method.[46] The enzyme column was very stable, and its half-life was 35 days at 45°C.

4.2.4   Lyase Reactions (asymmetric synthesis)

Several papers on applications of immobilized enzymes and immobilized microbial cells for asymmetric synthesis have been published. In this section, our studies on the production of L-aspartic acid, L-malic acid and urocanic acid by immobilized microbial cells are mainly described.

A.   Production of L-aspartic acid

L-Aspartic acid has been used as a medicine and food additive, and has been produced industrially from fumaric acid and ammonia by the fermentative or enzymic method, employing the catalytic action of aspartase (Eq. 4.16).

$$\underset{\text{fumaric acid}}{\text{HOOCCH=CHCOOH}}+\text{NH}_3 \underset{}{\overset{\text{aspartase}}{\rightleftharpoons}} \underset{\underset{\underset{\text{L-aspartic acid}}{\text{NH}_2}}{|}}{\text{HOOCCH}_2\text{CHCOOH}} \qquad (4.16)$$

For continuous production of L-aspartic acid using immobilized aspartase, the authors[10] investigated various immobilization methods for the *Escherichia coli* enzyme, as shown in Table 4.10.

Among the immobilization methods attempted, relatively active immobilized aspartase was obtained by the entrapping method using polyacrylamide gel. The stability of the immobilized enzyme column was investigated by operating it continuously for a long period. As shown in Fig. 4.17, the activity of the column decreased about 50% in the case of 1 M substrate solution after operation for about 30 days at 37°C.

For industrial application of this method, the enzyme has to be ex-

TABLE 4.10 Relationship between immobilization and yield of activity of aspartase[10]

Immobilization method and carrier	Native aspartase used (μmol/h)	Immobilized aspartase	
		Activity (μmol/h)	Yield of activity (%)
PHYSICAL ADSORPTION			
Silica gel	8400	10	0.1
Ca phosphate gel	4800	230	4.8
IONIC BINDING			
DEAE-cellulose	4800	276	5.8
ECTEOLA-cellulose	4800	35	0.7
TEAE-cellulose	4800	336	7.0
DEAE–Sephadex	4800	158	3.3
COVALENT BINDING			
Diazotized PAB-cellulose	700	3	0.4
CM-cellulose azide	700	0	0
ENTRAPPING			
Polyacrylamide gel	1440	417	29.0

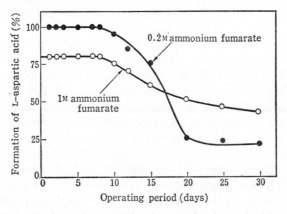

Fig. 4.17 Stability of an immobilized aspartase column.[10]
Aspartase was immobilized by the entrapping method using polyacrylamide gel.
A solution of 0.2 M or 1 M ammonium fumarate was continuously passed through the immobilized cell column at flow rate of $SV = 0.16$ at 37°C.

tracted from microbial cells, because it is intracellular. Also, the activity yield and the stability of the immobilized enzyme were not satisfactory. As the price of aspartic acid produced by the conventional method is relatively low, the use of the immobilized enzyme was not considered particularly advantageous for industrial purposes.

The authors considered that if microbial cells having the enzyme activity could be directly immobilized, these disadvantages might be overcome. Thus, the authors studied the immobilization of whole microbial cells.[95]

Immobilization of *Escherichia coli* cells having high aspartase activity was tested by the following methods: 1) entrapping in a polyacrylamide gel lattice prepared by using acrylamide monomer and a cross-linking agent such as N,N'-methylenebisacrylamide (BIS), 2) encapsulating in semipermeable polyurea produced from 2,4-toluene diisocyanate and hexamethylenediamine, and 3) cross-linking with a bifunctional reagent such as glutaraldehyde or 2,4-toluene diisocyanate. It was found that active immobilized cells were obtained by the entrapping method using polyacrylamide gel.

As cross-linking agents, in addition to BIS, N,N'-propylenebisacrylamide, diacrylamide dimethyl ether, 1,2-diacrylamide ethyleneglycol and N,N'-diallyl tartardiamide could be used, and the immobilized cells obtained by using these agents showed almost the same activity. However, BIS is preferred for industrial purposes, because it is commercially available in large amounts and at low cost. The chemical structure of polyacrylamide gel obtained by using BIS as a cross-linking agent is shown in Fig. 2.7 (p. 78), and an electron micrograph of the resulting immobilized *E. coli* cells is shown in Photo 2.2 (p. 78).

Although the pore size of this polyacrylamide gel, as described in chapter 2, can be varied by varying the concentrations of monomer and cross-linking agent used for polymerization, the average seems to be 10–40 Å under the conditions described previously (*see* p. 78). Accordingly, not only microbial cells, but also macromolecular enzyme proteins such as aspartase are retained in the gel lattice. However, low-molecular substrate and product can pass freely through the gel lattice.

An interesting phenomenon was observed with these immobilized cells. When freshly prepared immobilized *E. coli* cells were suspended at 37°C for 48 h in substrate solution, the activity increased about 10 times. As this phenomenon is very advantageous for continuous production of L-aspartic acid, the activation mechanism was investigated in detail. The following possibilities were considered; 1) adaptive formation of enzyme protein occurred in the presence of substrate, as the activation occurred when intact cells were incubated in the substrate solution, and 2) the apparent enzyme activity was elevated by an increase of membrane permeability for substrate and/or product due to autolysis of the cells in the gel lattice.

In order to determine whether the enzyme was adaptively formed by protein synthesis in the cells, the effect of chloramphenicol at a concentration that completely inhibited protein synthesis was investigated. It was found that the activation still occurred even in the presence of chloramphenicol, so that adaptive formation was ruled out. In this activation, oxygen uptake and glucose consumption of the immobilized cells decreased markedly. That is, the results suggested that the activation occurred as a result of autolysis of the cells in the gel lattice. This mechanism was confirmed by electron microscopy. As shown in Photo 4.3, in comparison with the cells immediately after immobilization, the photograph after activation shows that cell lysis had occurred. Accordingly, the elevation of the apparent enzyme activity was confirmed to be due to increased permeability caused by autolysis of *E. coli* cells in the gel lattice.

This activation, as shown in Fig. 4.18, occurred even when the intact cells were autolyzed in substrate solution, and the activity was similar to that of activated immobilized cells. When the autolyzed cells were homogenized, no significant elevation of aspartase activity was observed. On the other hand, when aspartase was immobilized after being extracted from *E. coli* cells, significant inactivation occurred during the immobilization process and the activity of the obtained immobilized enzyme was 1/20 of that of activated immobilized *E. coli* cells. Accordingly, in the case of an intracellular enzyme which is unstable when extracted, such as aspart-

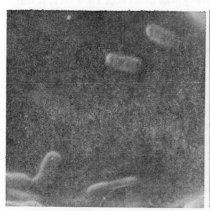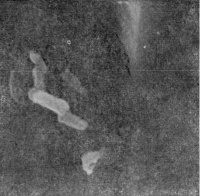

Photo 4.3 Electron micrographs of immobilized *E. coli* cells (×10,000).
Right, immobilized cells before activation;
Left, immobilized cells after activation.

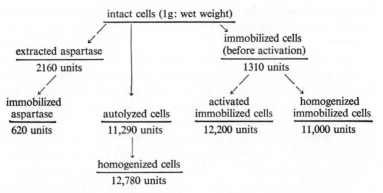

Fig. 4.18   Comparison of the aspartase activities of various enzyme preparations per unit of intact cells.
One unit is defined as the activity producing 1 μmol of L-aspartic acid at 37°C for 1 h.

ase, it appears that directly immobilized cells are preferable for industrial purposes.

To carry out a continuous enzyme reaction with activated immobilized cells, the enzyme properties were investigated. The optimum pH of the immobilized cells for this reaction varied considerably and the stability was increased somewhat in comparison with those of the intact cells. As it was reported that aspartase extracted from the cells was activated by Mn^{2+}, the effect of bivalent metal ions on the aspartase activity of the intact and immobilized cells was investigated. In this case, no activating effect of metal ions on the enzyme activity was observed. However, it was found that bivalent metal ions such as Mn^{2+}, Mg^{2+}, Ca^{2+} and others showed a stabilizing effect on aspartase activity of the immobilized cells.

In practice, when continuous reaction was carried out by passing a substrate solution (an aqueous solution of ammonium fumarate) through a column packed with immobilized cells, the activity of the immobilized cells decreased rapidly, as shown in Fig. 4.19.[96] However, when a substrate solution containing one of the bivalent metal ions described above was passed through the immobilized cell column its activity was retained without loss for a long period, as shown in Fig. 4.19. Further, the relationship between the stability of immobilized cells and temperature was investigated; the results are given in chapter 3 (Fig. 3.10). That is, when continuous reaction was carried out at 37°C, the immobilized cells were very stable and the half-life was more than 120 days.[9,11]

When an aqueous solution of ammonium fumarate containing Mg^{2+} was passed through a column packed with immobilized cells, the reaction in the column proceeded according to Eq. 4.16.

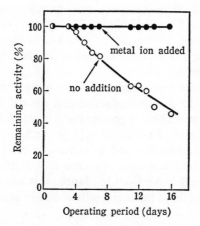

Fig. 4.19 Effect of metal ions on the stability of an immobilized *E. coli* cell column.[96] A solution of 1 M ammonium fumarate (containing 1 mM or no metal ions) was continuously passed through the immobilized cell column at a flow rate of $SV=0.5$ at 37°C. One of various metal ions such as Mg^{2+}, Mn^{2+} and Ca^{2+} was added to the solution of ammonium fumarate.

As the aspartase reaction is exothermic, the column used for industrial production of L-aspartic acid was designed as a multistage system with a radiator.

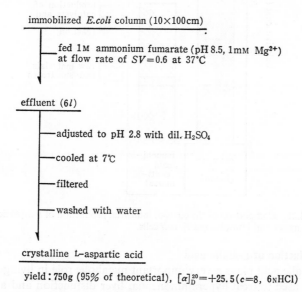

immobilized *E.coli* column (10×100cm)

    fed 1M ammonium fumarate (pH 8.5, 1mM Mg^{2+})
    at flow rate of $SV=0.6$ at 37°C

effluent (6*l*)

    —adjusted to pH 2.8 with dil. H_2SO_4

    —cooled at 7°C

    —filtered

    —washed with water

crystalline L-aspartic acid

    yield : 750g (95% of theoretical), $[\alpha]_D^{20}=+25.5$ ($c=8$, 6$_N$HCl)

Fig. 4.20 Procedure for the isolation and crystallization of L-aspartic acid.[96]

Contaminants such as microbial cells, proteins and so on are not present in the effluent from the column. Therefore, as shown in Fig. 4.20, L-aspartic acid of high purity can be obtained in high yield by a very simple procedure such as adjusting the pH of the effluent to the isoelectric point (pH 2.8–3.0) of the acid.[96]

Thus, in the continuous enzyme reaction using immobilized cells, as shown in Fig. 4.21, the overall production cost was reduced to about 60% of that of the conventional batch process using intact cells due to a marked increase of productivity of L-aspartic acid per unit of cells, reduction of labor costs due to automation and an increase in the yield of L-aspartic acid.[84] L-Aspartic acid has been produced industrially by this process since 1973 at Tanabe Seiyaku Co. Ltd., Japan. This was the first industrial application of immobilized microbial cells as a solid catalyst in the world.

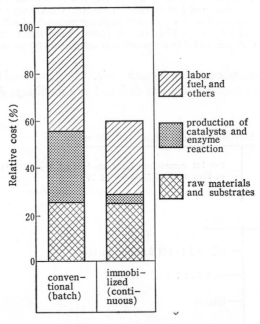

Fig. 4.21   Comparison of the costs of industrial production of L-aspartic acid using intact and immobilized *E. coli* cells.

B.   Production of L-malic acid

L-Malic acid plays a very important metabolic role in living cells and has been used to treat hyperammonemia, liver disfunction and as a component of amino acid infusions. Also, if it becomes possible to supply large amounts of L-malic acid at low cost, L-malic acid—the natural form —is expected to be used instead of DL-malic acid—the racemic form—in the food industry as a food additive.

Existing preparation methods for L-malic acid include isolation from natural fruit juice, the optical resolution of synthetic DL-form, fermentation and others. However, for industrial purposes, the acid has been produced from fumaric acid by the action of fumarase as shown in Eq. 4.17.

$$\text{HOOCCH} = \text{CHCOOH} + \text{H}_2\text{O} \xrightleftharpoons{\text{fumarase}} \underset{\underset{\text{L-malic acid}}{\overset{|}{\text{OH}}}}{\text{HOOCCH}_2\text{CHCOOH}}$$

fumaric acid (4.17)

The authors[12,97] considered that if microbial cells having high fumarase activity could be immobilized, malic acid of high purity might be efficiently produced in large amounts and at low cost by using the immobilized cells, as in the case of L-aspartic acid.

Several microorganisms having high fumarase activity were immobilized by the polyacrylamide gel method, and their activities were compared. *Brevibacterium ammoniagenes* was found to be the most active before and after immobilization.

When these immobilized cells were used for the production of L-malic acid from fumaric acid, succinic acid was formed as a by-product, and the separation of succinic acid from L-malic acid is technically very difficult. Accordingly, for the industrial production of pure L-malic acid in high yield, it was necessary to prevent the formation of succinic acid during the enzyme reaction. Thus, various treatments of intact or immobilized cells were performed. Heat treatment, autolysis and freezing-thawing were not effective for the prevention of succinic acid formation. However, treatment with substrate solution containing deoxycholic acid or bile acid, which is commonly used as a solubilizer of membrane-bound enzymes, was found to be very effective for the suppression of succinic acid formation. Further, this treatment also markedly enhanced the formation of L-malic acid. This effect was attributable to an increase of membrane permeability to the substrate and/or product due to bile acid treatment. Bile extract is suitable for industrial purposes as it is commercially available at a low price. Various conditions for treatment with bile extract were studied, and the most effective procedure was to stand the immobilized cells in 1 M sodium fumarate (pH 7.5) containing 0.3% bile extract at 37°C for 20 h. The yield of fumarase activity of immobilized cells obtained in this way is shown in Fig. 4.22.

Enzymic properties of the immobilized cells treated with bile extract were studied and compared with those of the native enzyme. The optimum temperature and optimum pH of the enzyme reaction were almost same as

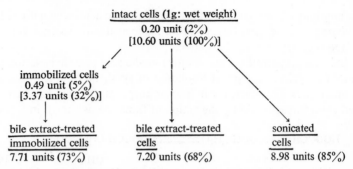

Fig. 4.22 Comparison of the fumarase activities of various enzyme preparations per unit of intact cells.[12]
One unit is defined as activity producing 1 mmol of L-malic acid at 37°C in 1 h.
Square brackets show activity in the presence of 0.02% cetylpyridium chloride; percentages are values relative to the activity (100%) of intact cells in the presence of cetylpyridium chloride.

those of the native enzyme, namely 60°C and pH 7.0–7.5, respectively. The equilibrium point of the enzyme reaction was scarcely affected by immobilization or by the reaction pH, but tended to shift toward L-malic acid formation with decreasing reaction temperature.

Conditions for the continuous enzyme reaction using a column packed with the immobilized cells were investigated. It was found that when 1 M sodium fumarate (pH 7.5) was passed through the column at 37°C at a space velocity of below 0.23 (residence time in the column: more than 4.3 h), the reaction reached equilibrium with about 80% conversion of fumaric acid to L-malic acid. Fumaric acid was removed from the effluent by acidification, and L-malic acid was separated in about 70% yield from fumaric acid.

For industrialization of this system, the authors carried out chemical engineering studies as described in sections 4.1.3 and 4.1.4, and designed a continuous reaction system similar to that for the production of L-aspartic acid using immobilized E. coli. L-Malic acid of high purity has been industrially produced at Tanabe Seiyaku Co. Ltd. by this process since 1974.

C. Production of urocanic acid

Urocanic acid is used as a sun-screening agent in the pharmaceutical and cosmetic fields, and has been produced from L-histidine by the action of L-histidine ammonia lyase as shown in Eq. 4.18.

Many kinds of microorganism having this enzyme activity have been reported. The authors[98] found that Achromobacter liquidum is the most suitable for industrial production, and the acid has been industrially pro-

$$\underset{\text{L-histidine}}{\underset{\overset{\displaystyle N}{\underset{\displaystyle H}{\diagdown C \diagup}}}{\overset{\displaystyle HC=\!\!=\!C\text{-}CH_2CHCOOH}{\underset{\displaystyle NH\ \ NH_2}{|\ \ \ \ \ |}}}} \xrightarrow[\text{ammonia lyase}]{\text{L-histidine}} \underset{\text{urocanic acid}}{\underset{\overset{\displaystyle N}{\underset{\displaystyle H}{\diagdown C \diagup}}}{\overset{\displaystyle HC=\!\!=\!C\text{-}CH=CHCOOH}{\underset{\displaystyle NH}{|\ \ \ \ \ |}}}} \ +\ NH_3$$

$$(4.18)$$

duced by using this microorganism at Tanabe Seiyaku Co. Ltd. To develop a more efficient process for industrial production of the acid, the authors[99] immobilized microbial cells having high L-histidine ammonia lyase activity by the polyacrylamide gel method, and investigated the continuous reaction using the immobilized cells packed into a column.

Several microorganisms having high enzyme activity were immobilized, and their activities were compared with those of the intact cells (Table 4.11). The yield of enzyme activity of immobilized cells varied considerably with the kind of microorganism. Of the tested microorganisms, *A. liquidum* showed the highest activity after immobilization and its activity yield was also high. On the other hand, in the case of *Micrococcus ureae*, the enzyme activity was almost lost on immobilization, although the enzyme activity of the intact cells was highest among the microorganisms tested. Thus, when microorganisms are immobilized in polyacrylamide gel, their activities may be markedly affected. Accordingly, even if the enzyme activity of intact cells is high, active immobilized cells are not always obtained in a good yield.

A. liquidum was considered to be the most suitable microorganism for the production of urocanic acid. However, *A. liquidum* was found to have urocanase activity, converting urocanic acid to imidazolone propionic acid. Thus, for effective production of urocanic acid, it was necessary to

TABLE 4.11   Enzyme activity of intact and immobilized cells of various microorganisms having L-histidine ammonia lyase activity[99]

Microorganism	Enzyme activity†		Yield of enzyme activity (%)
	Intact cells	Immobilized cells	
Achromobacter aceris	136	28	20.6
Achromobacter liquidum	1368	872	63.7
Agrobacterium radiobacter	928	480	51.7
Agrobacterium tumefaciens	740	368	49.7
Flavobacterium flavescens	1240	748	60.3
Micrococcus urea	4300	12	0.3
Sarcina lutea	560	188	33.6

† Urocanic acid μmol/h/g of wet cells.

prevent the action of this enzyme. Various procedures were investigated, and it was found that when *A. liquidum* was heated at 70°C for 30 min, the L-histidine ammonia lyase activity was not decreased but urocanase was completely inactivated.[98] Thus simple heat treatment at 70°C for 30 min before immobilization of the cells completely suppressed the conversion of urocanic acid to imidazole propionic acid by urocanase.

Such an unwanted side reaction is an important problem in cases where microbial cells having many kinds of enzyme are immobilized and used as solid catalysts. It is preferable to select microorganisms which have only the desired enzyme activity, without enzyme activity catalyzing side reactions. However, the selection of such microorganisms is difficult. Accordingly, some means to suppress undesired side reactions are necessary. It is also necessary to determine at the initial stage of the study whether such side reactions occur or not. Consequently, analysis of the enzyme reaction must be carried out not only in connection with the formation of product and the disappearance of substrate but also on the stoichiometry of the reaction.

The enzymic properties of immobilized *A. liquidum* cells were compared with those of intact cells. With regard to optimum pH and optimum temperature of the enzyme reaction, no difference between immobilized and intact cells was observed.

The permeability of substrate or product through the cell membrane increased on immobilization, as in the case of immobilized *P. putida*, and

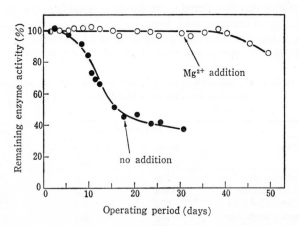

Fig. 4.23   Effect of metal ions on the stability of an immobilized *A. liquidum* column.[99]
A solution of 0.25 M L-histidine (pH 9.0, containing 1 mM or no Mg^{2+}) was continuously passed through the immobilized cell column at a flow rate of $SV = 0.1$ at 37°C.

the formation of urocanic acid proceeded without the addition of surfactant to the substrate solution.

When L-histidine solution (pH 9.0) containing Mg^{2+} was passed through a column packed with the immobilized *A. liquidum* cells, the amino acid was completely converted to urocanic acid. High purity urocanic acid was obtained from the effluent of the column in good yield, without recrystallization, by merely adjusting the pH of the effluent to around 4.7. This immobilized cell column was very stable, and when the continuous enzyme reaction was carried out in the presence of Mg^{2+}, as in the case of the immobilized *E. coli* column described previously, no apparent decrease of the activity was observed for 40 days, and its half-life was about 6 months. This stabilizing effect of Mg^{2+} and the stability of the immobilized *A. liquidum* column are shown in Fig. 4.23.

D. Other lyase reactions

In addition to our studies described above, other reactions of asymmetric synthesis have been reported. For instance, immobilized D-oxynitrilase was used for continuous conversion of benzaldehyde to D-(+)-mandelonitrile.[68] Namely, D-oxynitrilase immobilized by ionic binding to ECTEOLA-cellulose was packed into a column and a solution of benzaldehyde and hydrogen cyanide dissolved in 50% methanol was passed through it. Asymmetric synthesis (Eq. 4.19) proceeded in the column and

$$
\underset{\text{benzaldehyde}}{\text{C}_6\text{H}_5\text{CHO}} \;+\; \text{HCN} \;\xrightarrow{\;\text{D - oxynitrilase}\;}\; \underset{\text{D-(+)- mandelonitrile}}{\text{C}_6\text{H}_5\overset{\overset{\text{OH}}{|}}{\underset{\underset{\text{H}}{|}}{\text{C}}}\text{-CN}}
\tag{4.19}
$$

D-(+)-mandelonitrile of high optical purity (97% D- and 3% L-form) was obtained in high yield (95%). This enzyme can also be used for the conversion of aliphatic, aromatic and heterocyclic aldehydes to the corresponding nitriles in the presence of hydrogen cyanide.

4.2.5 Isomerization Reactions

Among isomerases, glucose isomerase has been extensively investigated for continuous enzyme reaction. Studies on the isomerization of glucose to fructose by immobilized glucose isomerase have been carried out for over 10 years to increase sweetness, and this continuous isomerization procedure has been carried out industrially.

So far, positive immobilization methods for this enzyme have been studied, and the enzyme has been immobilized by physical adsorption on alumina,[100] by ionic binding to DEAE-cellulose,[72,73] DEAE-Sephadex,[70] and ion-exchange resin,[101] by covalent binding to diazotized porous glass,[69] by cross-linking with glutaraldehyde[102] and by entrapping in polyacrylamide gel,[72,103] or cellulose triacetate.[104] Simley et al.[71] and Weetall et al.[105] built a pilot plant and carried out studies on the continuous isomerization of glucose using the immobilized enzyme.

Fig. 4.24 shows the flow diagram for continuous isomerization of glucose as carried out at Clinton Corn Processing Co., U.S.A. In this case, the enzyme was immobilized by ionic binding to DEAE-cellulose, and multishallow bed reactor were used. The half-life of the immobilized enzyme was reported to be "several hundred hours", and more than 500,000 tons of high-fructose corn syrup has been produced.[72,73]

At present, immobilized glucose isomerase and immobilized microbial cells, prepared by cross-linking with glutaraldehyde, from Novo Industri, Denmark, and immobilized glucose isomerase, prepared by ionic binding to DEAE–cellulose, from Standard Brands Co., U.S.A., have been used for industrial purposes.

In Japan many companies have been trying industrial application of immobilized enzyme and immobilized cell preparations for the isomerization of glucose.

For example, Takasaki and Kambayashi[106] reported a kind of immobilization of microbial cells by heat treatment alone without chemical treatment. That is, when whole cells of Streptomyces sp. having high glucose isomerase activity are heated at 60–85°C for 2–20 min, glucose isomerase is retained inside the cells. It was claimed that the enzyme fixed within the cells did not leak out from the cells, even if the cells were treated for a long time under conditions suitable for extracting the enzyme from the cells by autolysis.

These heat-treated cells have been used to produce high-fructose syrup industrially in a batch process.

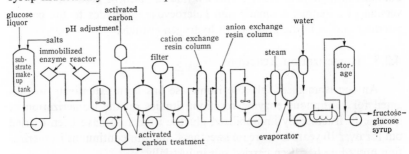

Fig. 4.24   Flow diagram for the continuous isomerization of glucose (corn syrup) using immobilized glucose isomerase (Clinton Corn Processing Co., U.S.A.)[72]

Further, conditions for the continuous production of high-fructose syrup were investigated by passing 40% glucose solution containing phosphate buffer (pH 8.0), 0.005 M magnesium sulfate and 0.001 M cobaltous chloride into a column packed with the heat-treated cells.[106] It was found that fructose syrup from glucose was obtained at an average isomerization ratio of 40% for 15 days. However, these heat-treated cells are not wholly satisfactory as regards enzyme activity, stability, and so on. Therefore, studies are being continued for the preparation of better stabilized immobilized cells.

Some of the remaining problems with the isomerization reaction are as follows: the pressure drops of immobilized enzyme and immobilized cell reactors increase during continuous operation, because glucose solution of relatively high concentration (40–45%) is used to facilitate the concentration procedure after the reaction. This high glucose concentration increases diffusion restriction and consequently decreases apparent activity. Accordingly, development of more efficient enzyme reactors is in progress.

Further, sodium sulfite is used for removing oxygen from the substrate solution, because oxygen has a marked effect on the stability and activity of the enzyme and causes the formation of colored by-products. In addition, cobaltous ions are added to the substrate solution to increase the stability and activity of the enzyme in many cases. These chemical substances are undesirable in the product and their addition causes increased expense in removing them subsequently.

If these problems can be resolved, industrial production of high-fructose syrup can be carried out advantageously using immobilized enzyme and immobilized cell preparations. Further development of immobilization methods and of enzyme reactors is expected to accelerate the change from conventional methods to immobilized enzyme systems or immobilized microbial cell systems.

4.2.6 Synthesis of Radioactive Compounds

Nowadays, radioactive compounds are widely used, and immobilized enzymes can be used advantageously for the synthesis of labeled compounds. In order to synthesize compounds containing isotopes of short half-life, such as ^{11}C (20 min) or ^{13}N (10 min) at high purity, immobilized enzymes are particularly useful.[107] For example, as shown in Eqs. 4.20 and 4.21, glutamate dehydrogenase or glutamate-pyruvate transaminase immobilized by covalent binding to aminoalkylated silica beads with glutaraldehyde were used for the synthesis of ^{13}N-L-glutamic acid and ^{13}N-L-alanine.

$$
\begin{array}{c}
\text{COOH} \\
| \\
\text{CH}_2 \\
| \\
\text{CH}_2 \\
| \\
\text{C=O} \\
| \\
\text{COOH}
\end{array}
+ {}^{13}\text{NH}_3 + \text{NADH}
\xrightleftharpoons{\text{glutamate dehydrogenase}}
\begin{array}{c}
\text{COOH} \\
| \\
\text{CH}_2 \\
| \\
\text{CH}_2 \\
| \\
\text{CH--}{}^{13}\text{NH}_2 \\
| \\
\text{COOH}
\end{array}
+ \text{NAD}
\qquad (4.20)
$$

$$
\begin{array}{c}
\text{COOH} \\
| \\
\text{CH}_2 \\
| \\
\text{CH}_2 \\
| \\
\text{CH--}{}^{13}\text{NH}_2 \\
| \\
\text{COOH}
\end{array}
+
\begin{array}{c}
\text{CH}_3 \\
| \\
\text{C=O} \\
| \\
\text{COOH}
\end{array}
\xrightleftharpoons{\text{glutamate-pyruvate transaminase}}
\begin{array}{c}
\text{COOH} \\
| \\
\text{CH}_2 \\
| \\
\text{CH}_2 \\
| \\
\text{C=O} \\
| \\
\text{COOH}
\end{array}
+
\begin{array}{c}
\text{CH}_3 \\
| \\
\text{CH--}{}^{13}\text{NH}_2 \\
| \\
\text{COOH}
\end{array}
\qquad (4.21)
$$

4.2.7   Multiple Enzyme Reactions

Continuous enzyme reactions using immobilized enzymes and immobilized microbial cells as described in the previous sections are carried out by the action of a single enzyme. However, in the production of many useful compounds by the action of microorganisms, especially by fermentative methods, the compounds are not produced by the action of a single enzyme but by the actions of several kinds of enzymes, that is, by a multienzyme system. Attempts have therefore been made to produce various compounds using two or more immobilized enzymes.

For example, the preparation of nicotinamide adenine dinucleotide (NAD) was investigated using two kinds of immobilized enzymes.[108] NAD was prepared from nicotinamide mononucleotide (NMN) and ATP according to Eq. 4.22.

$$
\text{NMN} + \text{ATP} \xrightleftharpoons{\text{NAD pyrophosphorylase}} \underset{\text{pyrophosphate}}{\text{NAD} + \text{PP}i}
$$

$$
\text{PP}i + \text{H}_2\text{O} \xrightleftharpoons{\text{inorganic pyrophosphatase}} \underset{\text{phosphate}}{\text{P}i}
\qquad (4.22)
$$

As the NAD pyrophosphorylase catalyzing this reaction is inhibited by the inorganic pyrophosphate produced, this should be removed from the reaction mixture for efficient formation of NAD. Thus, NAD pyrophosphorylase was immobilized by physical adsorption on hydroxylapatite and inorganic pyrophosphatase by ionic binding to DEAE-cellulose, and both immobilized enzymes were packed into the same column. An aqueous solution of NMN and ATP as substrates was passed through the column, and NAD was efficiently obtained from the column effluent.[108]

Recently, an interesting report on the preparation of L-lysine using two kinds of immobilized enzymes has been published by Toyo Rayon Industries Inc., Japan.[109] Cyclohexane, a by=product of nylon production, was used as a starting material, and was converted to DL-α-amino–ε-caprolactam (**4**) by a usual synthetic method. L-α-Amino–ε-caprolactam hydrolase and α-amino–ε-caprolactam racemase were immobilized by ionic binding to anion exchange polysaccharides such as DEAE–Sephadex, and both immobilized enzymes were allowed to react with DL-α-amino–ε-caprolactam at the same time. As shown in Eq. 4.23, L-α-amino–ε-caprolactam was hydrolyzed to give L-lysine, and the remaining D-α-amino–ε-caprolactam was racemized to the DL-form by the action of racemase and again hydrolyzed to L-lysine.

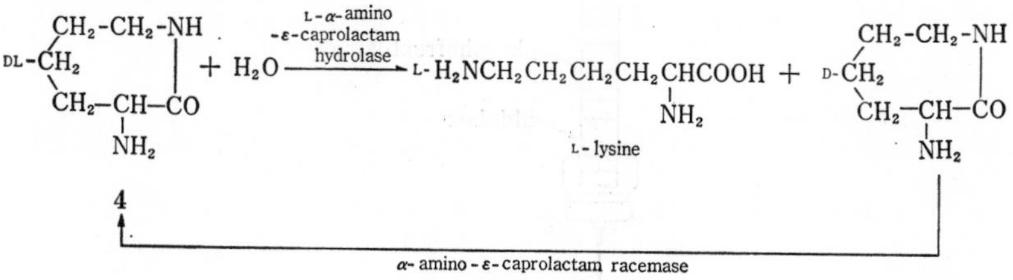

$$(4.23)$$

Repeating the above reactions, essentially all the DL-form was finally converted to L-lysine. The reaction can be carried out by the column or batch method.

As well as these systems, four kinds of glycolytic enzymes catalyzing Eq. 4.24 were immobilized separately by the entrapping method using

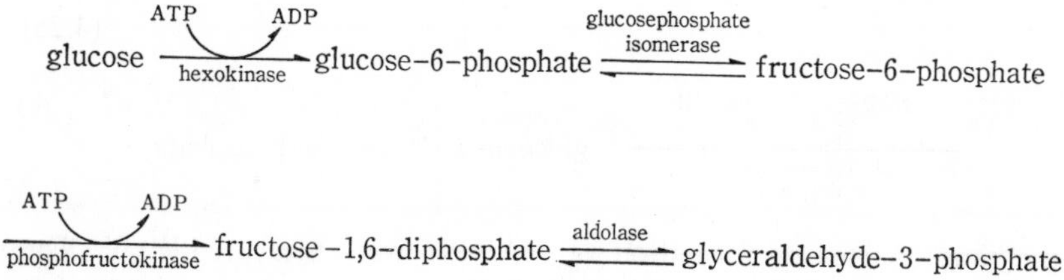

$$(4.24)$$

polyacrylamide gel, and the immobilized enzymes were packed into a column as shown in Fig. 4.25. When a solution containing glucose, ATP and magnesium ions was passed through the column glyceraldehyde–3-phosphate and ADP were obtained from the effluent.[110]

Further, two kinds of hexokinase (HK) and glucose-6-phosphate dehydrogenase (G-6-PDH) catalyzing the reaction shown in Eq. 4.25 were immobilized at various ratios of activities on the same carrier, and

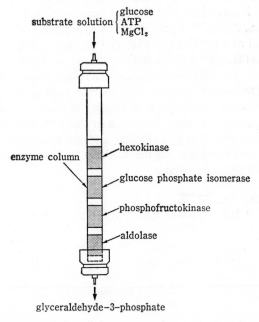

Fig. 4.25   Column for continuous enzyme reaction using four kinds of immobilized enzymes.[110]

$$
\text{glucose} \xrightarrow[\text{hexokinase}]{\overset{\text{ATP} \qquad \text{ADP}}{\curvearrowright}} \text{glucose-6-phosphate}
$$

$$(4.25)$$

$$
\xrightarrow[\substack{\text{glucose-6-phosphate} \\ \text{dehydrogenase}}]{\overset{\text{NADP}^+ \qquad \text{NADPH}}{\curvearrowright}} \text{glucono-}\delta\text{-lactone-6-phosphate}
$$

the time courses of formation of NADPH were measured.[111~113] The formation of NADPH using immobilized enzymes was greater than that using free enzymes in all cases. In a consecutive reaction a time lag exists before formation of the final product, although the lag is less in the case of the immobilized enzyme system than in the case of the soluble enzyme system. These results can be interpreted as follows. In the immobilized enzyme system, glucose-6-phosphate (G-6-P) produced by the first enzyme reaction is immediately oxidized by the second enzyme, G-6-PDH, at a location close to the first one. In other words, the produced G-6-P can not easily pass to the outer solution by diffusion so the G-6-P is easily oxidized by surrounding G-6-PDH since it is present at an apparently high concentration (even if the average concentration in the whole reaction system is low, the effective concentration is higher in the inner region of the carrier). This is a good example where diffusion resistance is efficiently utilized, and it suggests that the location of enzymes within cells significantly affects the enzyme reaction rate.

Further, when the activity ratio of HK and G-6-PDH is small, the difference between immobilized and free enzymes becomes great (Fig. 4.26 (b) and (c)), while when it is high, the difference becomes small. In the latter case, it is considered that as the concentration of G-6-P within the carrier becomes very high due to the high activity of HK, the difference between the two systems becomes less apparent, because rather large amounts of G-6-P are released into the outer solution against the diffusion resistance. That is, when the activity ratio of HK and G-6-PDH is great, the diffusion resistance is not utilized efficiently.

In addition, theoretical analysis of such consecutive reactions by immobilized multi-enzyme systems has been extensively studied.[114,115]

As well as the multi-enzyme systems described above, aldolase and glyceraldehyde–3-phosphate dehydrogenase were immobilized separately by covalent binding to AE-cellulose with glutaraldehyde, packed separately into different columns, and the reaction shown in Eq. 4.26 was carried out.[116] Into the first column packed with immobilized aldolase, a solution of fructose-1,6-diphosphate and NAD$^+$ was charged, and the effluent was passed through the column packed with immobilized glyceraldehyde-3-phosphate dehydrogenase to give NADH.

dihydroxyacetone phosphate

fructose–1,6–diphosphate $\xrightarrow[\text{aldolase}]{}$ glyceraldehyde–3–phosphate

NAD$^+$ \quad NADH

glyceraldehyde–3–phosphate dehydrogenase $\xrightarrow{}$ 1,3–diphosphoglyceric acid

$$(4.26)$$

When compounds are produced by multi-enzyme systems, reaction systems conjugated with an energy-generating system (regeneration system for ATP) and/or a redox system participate in many cases. Recently, studies on the immobilization of multi-enzyme systems combined with such conjugated systems have been carried out. Bioreactors combining these systems have been of particular interest in areas involving biochemical reactions.

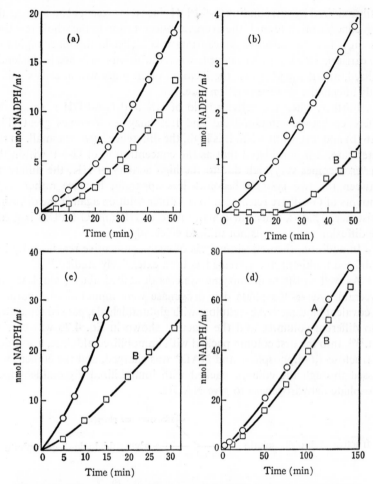

Fig. 4.26   Time courses of formation of NADPH from glucose by various hexokinase-glucose-6-phosphate dehydrogenase systems.[113)]

A: HK-G-6-PDH system immobilized simultaneously on a carrier.

B: Soluble enzyme system corresponding to A.

(a)   Carrier, Sepharose (covalent binding);
enzyme activity, HK = 1.1 nmol G-6-P/min/ml, G-6-PDH = 1.3 nmol NADPH/min/ml.

(b)   Carrier, copolymer of acrylamide and acrylic acid (covalent binding);
enzyme activity, HK = 0.25 nmol G-6-P/min/ml, G-6-DPH = 0.94 nmol NADPH/min/ml.

(c)   Carrier, polyacrylamide gel (entrapping);
enzyme activity, HK = 1.3 nmol G-6-P/min/ml, G-6-DPH = 14.0 nmol NADPH/min/ml.

(d)   Carrier, Sepharose (covalent binding);
enzyme activity, HK = 2.4 nmol G-6-P/min/ml; G-6-DPH = 1.1 nmol NADPH/min/ml.

For example, as shown in Fig. 4.27, Kamen and Kaplan,[117] of the University of California, U.S.A., have regenerated co-enzyme by coupling with an immobilized reduction system and carried out the reduction of steroids. Further, as shown in Fig. 4.28, immobilizations of complicated redox and ATP-regenerating systems have been studied.

As shown in Fig. 4.29, Wang and co-workers[118] at MIT, U.S.A., have studied an enzymic ATP-regenerating system using acetylphosphate, a donor of phosphate prepared by chemical synthesis. Using this regeneration system, synthesis of gramicidin S has been studied.

On the other hand, studies on the application of immobilized microbial cells for reactions in which several enzymes participate, especially fermentative processes, have recently been carried out.[119] *Corynebacterium glutamicum*, a well-known glutamic acid-producing microorganism, was immobilized by the polyacrylamide gel method using acrylamide and BIS. This immobilized *C. glutamicum* was suspended in a solution containing glucose as a carbone source, inorganic ammonium salts (ammonium sulfate and ammonium phosphate) as a nitrogen source and several kinds of metal ions, and glutamic acid fermentation was carried out with shaking by a batch method. As shown in Fig. 4.30, L-glutamic acid was accumulated in the reaction mixture, and the concentration of the acid reached 15 g/l after incubation for 144 h. Fig. 4.30 shows that the productivity of fresh immobilized cells for L-glutamic acid was higher than that of reused immobilized cells.

These results suggest the possibility that some of the conventional fermentative methods can be continuously carried out by using columns

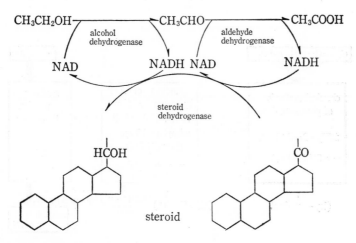

Fig. 4.27   Bioreactor process for oxidation and reduction.

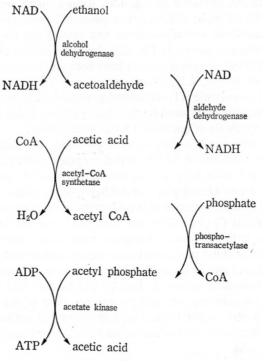

Fig. 4.28 Bioreactor process for oxidation-reduction and ATP—generating systems.

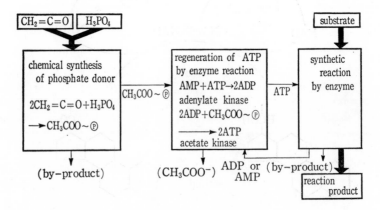

Fig. 4.29 Bioreactor for regeneration of ATP.[118]

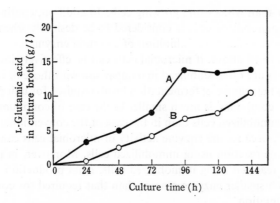

Fig. 4.30 Glutamic acid fermentation using immobilized cells.[119]
A, Freshly prepared immobilized cells; B, previously used immobilized cells.

packed with immobilized cells. However, when reactions requiring oxygen are carried out by the column method, the supply of oxygen into the column is a major problem, though it is expected to be overcome by the studies or reactors.

In addition, *Brevibacterium ammoniagenes*, used for the fermentative production of coenzyme A, was immobilized by the polyacrylamide gel method, and continuous production of coenzyme A was investigated using the immobilized cells.[120] When a solution containing pantothenic acid, cysteine, ATP and magnesium sulfate in potassium phosphate buffer (pH 6.5) was passed through a column packed with the immobilized cells, coenzyme A was formed in the effluent. Under the optimum conditions, it was reported that 500 μg of coenzyme A per ml of the effluent was accumulated; the activity of the immobilized cells decreased to about 50% after continuous operation for 1 week.

In addition to these examples, *Arthrobacter oxydans* immobilized in polyacrylamide gel was used for the production of lactic acid from 1,3-propanediol in both batch and column processes.[121]

Further, more recently, an interesting study on the continuous production of beer using immobilized microbial cells has appeared.[122] Yeast bound to porous bricks was packed into two columns; one column was used to convert wort into beer and the other column was used for the maturation of the beer. The quality of the beer produced is said to be not outstanding.

These results suggest the possibility that some of the conventional fermentative methods can be replaced in the future by continuous enzyme reactions using immobilized enzymes or immobilized microbial cells. In the case of immobilized cells particularly, control of the enzyme reaction is much easier than when using growing cells. In addition, continuous reaction using immobilized cells is considered to be desirable when a desired compound shows feedback inhibition of a certain enzyme.

As described above, if microbial cells can be effectively immobilized, their enzymes are often stabilized in comparison with those of intact cells. In addition, the volume of fermentation broth required for unit production of a desired compound is much smaller in the case of continuous production using immobilized cells than in the case of the conventional fermentative method, because the enzyme activity of microbial cells can be reused in the enzyme reaction using immobilized cells. Moreover, in the case of continuous reaction using immobilized cells, mass production is possible using a much smaller enzyme reactor than that required for conventional batch fermentation.

Further, as regards problems with the treatment of fermentative wastes to prevent water pollution, the cost of energy, etc., this technique is very promising, and major developments are expected in the future.

Besides these applications, application to artificial organs and attempts to substitute for certain physiological functions by employing microencapsulated multi-enzymes have also been made.[123] For example, carbamate kinase and ornithine transcarbamylase have been microencapsulated by the liquid drying method using ethyl cellulose. These enzymes catalyze the reaction shown in Eq. 4.27, and produce L-citrulline from ammonia, carbon dioxide, ATP and ornithine.

$$ATP + NH_3 + CO_2 + H_2O \xrightarrow{\text{carbamate kinase}} ADP + Pi + \underset{\underset{O}{\overset{||}{}}}{\overset{\overset{OH}{\overset{|}{}}}{HOPOCONH_2}}$$

$$\text{carbamyl phosphate}$$

$$\text{carbamyl phosphate} + H_2NCH_2CH_2CH_2\underset{\underset{NH_2}{|}}{CHCOOH}$$

$$\text{L-ornithine}$$

$$\xrightarrow[\text{transcarbamylase}]{\text{L-ornithine}} H_2NCONHCH_2CH_2CH_2\underset{\underset{NH_2}{|}}{CHCOOH} + Pi$$

$$\text{L-citrulline}$$

$$(4.27)$$

Many enzymes exist *in vivo* bound to cell membranes and/or particles. Immobilized enzymes can thus be used to elucidate reaction mechanisms *in vivo* as models of such bound enzyme systems. This is described in detail later (section 4.3.3).

4.3 ANALYTICAL APPLICATIONS

Recently, attempts to utilize the special characteristics of immobilized enzymes for analysis have been reported. These attempts can be roughly classified as follows.

1) Development of automated analysis procedures by combining immobilized enzymes with various analytical instruments, including the development of biochemical electrodes.

2) Structural analysis of polymers such as proteins, nucleic acids, and so on.

3) Analysis of complex enzyme reaction mechanisms or of enzyme function.

Various books have been published in this field, in addition to articles cited here: those of Weetall[124] and Guilbault[125-128] of the U.S.A. on applications of enzymes, including immobilized enzymes, for analysis, and of Gough and Andrade[129] on enzyme electrodes, can be recommended.

4.3.1 Automated Analysis and Biochemical Electrodes

Analytical methods involving combinations of immobilized enzymes with analytical instruments are finding wide application in the fields of chemistry and medicine.

A. Automated analysis

By combining immobilized enzymes with spectrophotometers, fluorometers, micro-calorimeters and so forth, automation of various chemical and clinical analysis procedures has been attempted. Assay methods carried out by means of immobilized enzymes combined with a spectrophotometer are summarized in Table 4.12.

As shown in the table, alcohol dehydrogenase, lactate dehydrogenase, and malate dehydrogenase, which have NAD as a coenzyme, were immobilized by the carrier cross-linking method using glutaraldehyde and nylon tube, and automatic determinations of ethanol, lactic acid, and malic acid were carried out by following the formation of NADH at 340 nm.[130] The ranges of detectable concentration were 10–100 mM for ethanol, 0.04–0.2 mM for lactic acid, and 0.02–0.16 mM for malic acid. These nylon tubes binding enzymes could be used to measure at least 1000 samples.

The assay of lactic acid or glucose has been automatically carried out as shown in Fig. 4.31 by using a stable enzyme column packed with immobilized lactate dehydrogenase or glucose oxidase prepared by the poly-

TABLE 4.12 Automated analysis using immobilized enzymes in combination with a spectrophotometer

Enzyme	Immobilization method	Substance detected	Reference
Alcohol dehydrogenase	carrier cross-linking (nylon tube+ glutaraldehyde)	ethanol	130
Lactate dehydrogenase	entrapping (polyacrylamide gel)	lactic acid	131
	carrier cross-linking (nylon tube+ glutaraldehyde)	lactic acid	130
Malate dehydrogenase	carrier cross-linking (nylon tube+ glutaraldehyde)	L-malic acid	130
Glucose oxidase	entrapping (polyacrylamide gel)	glucose	131
	carrier cross-linking (nylon tube+ glutaraldehyde)	glucose	132, 133
D-Amino acid oxidase	peptide binding (CNBr-activated Sepharose)	D-amino acid	134
Urease	diazo binding (amino acid copolymer)	urea (urine, serum)	135
	carrier cross-linking (nylon tube+ glutaraldehyde)	urea	132
Invertase+ glucose oxidase	carrier cross-linking (nylon tube + glutaraldehyde)	sucrose	136
Glucoamylase+ glucose oxidase	carrier cross-linking (nylon tube + glutaraldehyde)	maltose	136
Lactase+ glucose oxidase	carrier cross-linking (nylon tube + glutaraldehyde)	lactose	136
Aspartate amino-transferase	peptide binding (CNBr-activated Sepharose)	L-aspartic acid	137
Aspartate amino-transferase+ malate amino-transferase	peptide binding (CNBr-activated Sepharose)	L-aspartic acid	137
Tryptophanase	peptide binding (CNBr-activated Sepharose)	L-tryptophan	138
Tryptophanase+ lactate dehydro-genase	peptide binding (CNBr-activated Sepharose)	L-tryptophan	138
Nitrate reductase	diazo binding (Porous glass)	nitrate ions	139

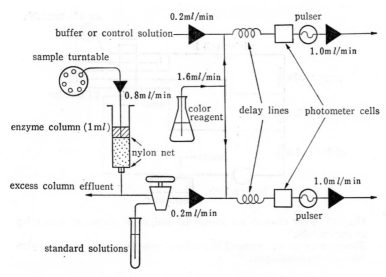

Fig. 4.31 Flow diagram for glucose or lactic acid determination.[131]

acrylamide gel method.[131] When a substrate solution was passed through the column at a constant flow rate, the reaction shown in Eq. 4.29 or 4.31 occurred, and colorimetric determinations were performed by measuring the decrease or increase of the color developed by the reaction shown in Eq. 4.30 or 4.32 on the subsequent addition of 2,6-dichlorophenolindophenol (DPI) to the reaction product.

1) $$\text{lactic acid} + \text{NAD} \underset{}{\overset{\text{lactate dehydrogenase}}{\rightleftharpoons}} \text{pyruvic acid} + \text{NADH}_2 \qquad (4.29)$$

$$\underset{\text{(blue)}}{\text{NADH}_2 + \text{oxidized DPI}} \xrightarrow{\text{phenazine methosulfate}} \text{NAD} + \underset{\text{(colorless)}}{\text{reduced DPI}} \qquad (4.30)$$

2) $$\text{glucose} + \text{O}_2 \xrightarrow{\text{glucose oxidase}} \text{gluconic acid} + \text{H}_2\text{O}_2 \qquad (4.31)$$

$$\underset{\text{(colorless)}}{\text{H}_2\text{O}_2 + \text{reduced DPI}} \xrightarrow{\text{peroxidase}} \text{H}_2\text{O} + \underset{\text{(blue)}}{\text{oxidized DPI}} \qquad (4.32)$$

Analytical methods combining immobilized enzymes with an autoanalyzer have also been reported.[130,132] For instance, automatic analysis of glucose or urea was carried out in the flow system shown in Fig. 4.32 by using glucose oxidase or urease immobilized on a nylon tube with glutaraldehyde. Measurement of 150 samples per day was claimed to be possible over 30 days.[132]

The authors showed that D-amino acid oxidase immobilized on CNBr–activated Sepharose could be efficiently used for the determination

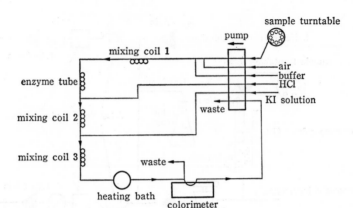

Fig. 4.32   Flow diagram for automated analysis of glucose or urea using an enzyme tube.[132)]
The enzyme tube was prepared by linking glucose oxidase or urease to nylon tube with glutaraldehyde.

of D-amino acids and of the optical purity of L-amino acids.[134)] It was reported that the determination of L-tryptophan could be easily carried out by using a multi-enzyme system simultaneously binding tryptophanase and lactate dehydrogenase to CNBr–activated Sepharose.[138)]

A column packed with urease immobilized on diazotized amino acid copolymer has been used for the determination of urea in urine or serum, and also for the removal of urea from body fluids.[135)]

Immobilized enzymes have also been used in an interesting analytical method based on the heat of chemical reaction.[140~142)] That is, trypsin was immobilized on polyacrylamide gel, packed into a column equipped with a micro-calorimeter, and the concentrations of samples were measured. In this case, the remaining activity of the column after operation for 4 weeks was 85%, using benzoyl–L-arginine ethyl ester as a substrate.[140)]

The determination of enzyme inhibitor by means of fluorometric analysis was reported.[143)] Cholinesterase entrapped in a starch matrix was used for the continuous detection of anti-cholinesterase compounds in water or air. This analysis is based on the use of a compound emitting fluorescence on enzyme reaction as a substrate; the presence of an anti-cholinesterase compound suppresses the fluorescence emission due to inhibition of the enzyme reaction.

B.   Biochemical electrodes

The use of biochemical electrodes having bio-specificity has been attracting considerable attention recently for the checking or control of concentrations of metabolites in body fluids. This bio-specificity is based on enzyme, and immobilized enzymes are used in many cases. These electrodes are suitable for the measurement of substrates, coenzymes, or inhibitors of an enzyme, and are called "enzyme electrodes" or "microbial electrodes" when microbial cells are used.

The enzyme electrode consists of an enzyme section reacting specifically with the test compound and an electrode for detecting changes in the charge of the compound. A glass electrode is most commonly used, and is suitable for the detection of changes in concentration of hydrogen ions. More than 20 kinds of specific ion electrodes have been prepared. Among these electrodes, monovalent cation electrodes for ammonium, potassium, sodium and cyanide ions, and ion-selective electrodes having specificity for phosphate are suitable for analysis of enzyme reactions. Oxygen or carbon dioxide electrodes for the measurement of oxygen or carbon dioxide dissolved in solutions are also useful.

An enzyme electrode can be made by immobilizing the enzyme on these electrodes or by covering the electrode with previously prepared immobilized enzyme.

Updike and Hicks[144] first prepared an enzyme electrode and used it for quantitative analysis. This enzyme electrode, shown in Fig. 4.33, con-

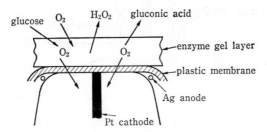

Fig. 4.33   Principle of the enzyme electrode for glucose measurement.[144]

sisted of an enzyme membrane (20–50 μ thick) of glucose oxidase immobilized by entrapping in polyacrylamide gel, a synthetic polymer membrane (e.g., Teflon) capable of passing oxygen, and an oxygen electrode. When glucose in the sample solution diffuses into the enzyme membrane, oxygen is consumed by the enzyme reaction (Eq. 4.33) and the

$$\text{glucose} + O_2 \xrightarrow{\text{glucose oxidase}} \text{gluconic acid} + H_2O_2 \qquad (4.33)$$

supply of oxygen to the electrode decreases. By measuring the oxygen concentration with the oxygen electrode, the amount of glucose can be determined. There is a linear relationship between the amount of glucose and the decrease of oxygen level over a certain range, and the measurement error is low, but some difficulty remains regarding temperature sensitivity. Application for automatic and continuous determination of glucose by directly inserting the enzyme electrode into a living body has been also studied. This system requires no reagent, i.e., so-called "reagentless analysis" can be performed.

Humphrey and his co-workers also reported the determination of glucose in solution by using such an enzyme electrode.[145] In this case, the sample was passed through a column packed with glucose oxidase immobilized on diazotized porous glass, and by measuring the concentration of oxygen in the effluent with an oxygen electrode, determination of glucose at a level of $10^{-5}-10^{-4}$ M was possible. Enzyme electrodes for this glucose determination are now available commercially.

As well as enzyme electrodes made by combining various electrodes with immobilized enzymes prepared by entrapping or covalent binding, the device shown in Fig. 4.34 has been used as an enzyme electrode. An

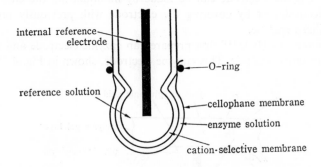

Fig. 4.34   L-Amino acid oxidase electrode using a cellophane membrane.

electrode is covered with a semipermeable membrane (ultrafiltration membrane) such as cellophane, and enzyme is entrapped in the space between the electrode and the membrane. Only low-molecular-weight substrate can pass through the cellophane membrane to come into contact with the enzyme, and the resulting reaction product is detected with the electrode.

Enzyme electrodes using immobilized enzymes which have so far been reported are listed in Table 4.13.

As well as the so-called enzyme electrodes, a monitoring system for reaction product in the effluent from an immobilized enzyme column was reported.[161] When fermentation broth containing penicillin was passed through a column packed with immobilized penicillinase, the reaction shown in Eq. 4.34 occurred and penicilloic acid was produced in the effluent. By detecting the acid with an electrode, measurement of penicillin

TABLE 4.13 Enzyme electrodes using immobilized enzymes

Immobilized enzyme	Substance detected	Electrode	Range of measurement (M)	Reference
POLYACRYLAMIDE GEL METHOD				
L-Amino acid oxidase	L-amino acid	monovalent cation electrode	10^{-4}–10^{-2}	146
L-Amino acid oxidase	L-phenyl-alanine	ammonium electrode	5×10^{-5}–10^{-2}	147
L-Amino acid oxidase and Peroxidase	L-phenyl-alanine	iodide electrode	5×10^{-5}–10^{-3}	147
D-Amino acid oxidase	D-amino acid	monovalent cation electrode	10^{-5}–10^{-2}	148, 149
Glucose oxidase	glucose	oxygen electrode	10^{-5}–10^{-4}	144, 150
Urease	urea	monovalent cation electrode	5×10^{-5}–1.6×10^{-1}	151
Penicillinase	penicillin	glass electrode	10^{-4}–5×10^{-2}	152
β-Glucosidase	amygdalin	cyanide electrode	10^{-5}–10^{-2}	153
Asparaginase	asparagine	monovalent cation electrode	5×10^{-2}–10^{-2}	149
COLLAGEN MEMBRANE METHOD				
Catalase	H_2O_2	oxygen electrode	10^{-4}–1.5×10^{-3}	154
Alcohol dehydrogenase	ethanol	platinum electrode	5×10^{-2}–5×10^{-1}	155
Lactate dehydrogenase	lactic acid	platinum electrode	7×10^{-3}–7×10^{-2}	155
Cholesterol oxidase	cholesterol	oxygen electrode	2×10^{-5}–1×10^{-4}	156
Invertase Mutarotase Glucose oxidase	sucrose	oxygen electrode	10^{-3}–10^{-2}	157
ULTRAFILTRATION MEMBRANE METHOD				
Urease	urea	carbon dioxide electrode	10^{-4}–10^{-1}	158
L-Tyrosine decarboxylase	L-tyrosine	carbon dioxide electrode	10^{-4}–10^{-1}	158
Glutaminase	glutamine	monovalent cation electrode	10^{-4}–10^{-1}	148
Glutamate dehydrogenase	L-glutamic acid	carbon dioxide electrode	2.8×10^{-5}–6.8×10^{-4}	159
Glutamate dehydrogenase	L-glutamic acid	cation electrode	10^{-4}–10^{-3}	160
Lactate dehydrogenase	pyruvate	cation electrode	2.5×10^{-5}–8×10^{-4}	160

$$R-CO-NH-CH-CH \underset{CO-N}{\overset{S}{\underset{\underset{CH-COOH}{|}}{\swarrow}}} C \overset{CH_3}{\underset{CH_3}{\diagdown}} + H_2O \xrightarrow{\text{penicillinase}}$$

$$R-CO-NH \underset{\underset{COOH}{|}}{——} CH \underset{\underset{NH-CH-COOH}{|}}{\overset{S}{\diagup}} C \overset{CH_3}{\underset{CH_3}{\diagdown}} \qquad (3.34)$$

at a level of 0.8×10^{-4} to 5×10^{-4} M was reported to be possible.

A microorganism, *Streptococcus faecalis*, was entrapped on an ammonia-sensitive electrode and measurement of arginine was carried out.[162] In this case, microorganisms metabolizing arginine to give ammonia were immobilized on the electrode by using a cellophane membrane. The sensitivity of this method is very high, and determination of $10^{-2}M–10^{-5}M$ arginine was achieved.

Recently, a BOD sensor combining an oxygen electrode and a membrane carrying immobilized microorganisms prepared by the polyacrylamide gel method has been reported.[163] Using this sensor, BOD values of wastes from alcohol plant, fermentation processes, slaughterhouses, and so on, can be measured within 15 min. This BOD sensor is stable, and constant current values were still obtained after usage for 10 days.

C. Simple assay and clinical test methods

In order to measure the amount of glucose in urine, a filter-paper strip soaked with buffer solution containing glucose oxidase and peroxidase, and an appropriate color reagent is widely used, and this is also an application of immobilized enzymes. Other measurements of specific substances have been attempted, especially for clinical tests.

For example, a simple and rapid measurement procedure for hydrogen peroxide was developed by using a paper strip of immobilized enzyme.[164] That is, peroxidase was covalently attached to a CM–cellulose strip in the presence of N,N'-dicyclohexyl-carbodiimide, and color was developed by the oxidation of benzidine soaked in the strip with oxygen formed by the enzyme action. Although hydrogen peroxide is detectable to a level of 10^{-6} M, by using a fluorescent substance as a hydrogen donor, as little as 5×10^{-11} M can be detected.

For clinical tests, a strip of DEAE–cellulose attaching lactase with Procion brilliant orange has been used to measure lactose in excrement.[165] This test is based on the fact that some tumor-bearing patients excrete considerable amounts of disaccharides such as lactose or sucrose.

4.3.2   Structural Analysis of Biopolymers

As a procedure to study the primary structure, amino acid composition or base composition of biopolymers such as proteins, nucleic acids, and so on, limited hydrolysis by proteolytic enzymes, carboxypeptidase, leucine aminopeptidase, nuclease, phosphodiesterase, or phosphomonoesterase has been carried out. However, in the conventional method using soluble enzyme, a relatively complicated treatment including the removal of the enzyme protein after the reaction is necessary for the preparation of analytical samples. On the other hand, if immobilized enzymes can be used, analytical samples can be prepared easily, simply by removing the immobilized enzymes by centrifugation or filtration. Structural analysis of proteins or nucleic acids using immobilized enzymes has already been carried out.

A.   Proteins and polypeptides

Immobilized enzymes have been used for the limited hydrolysis of proteins or determination of amino acid sequences of polypeptides as follows.

By using trypsin immobilized by diazo coupling with amino acid copolymer, the amino acid sequences of proteins or polypeptides have been determined.[166] Similarly, the mechanism of degradation of immunogloblin was studied using papain immobilized by diazo coupling with leucine and *p*-aminophenylalanine copolymer.[168]

The pronase from *Streptomyces griseus* is known to act on many kinds of peptide bonds and to have low substrate specificity. This enzyme, immobilized by covalent binding with porous glass, has been used for the isolation of various molecular sizes of peptides.[168] Leucine aminopeptidase and aminopeptidase M immobilized by the same method have been used for the complete hydrolysis of proteins, determination of amino acid sequences or measurements of optical purity of bound amino acid.[169]

Further, immobilized trypsin, chymotrypsin, prolidase, and aminopeptidase prepared by using CNBr–activated Sepharose have been utilized for the determination of amino acid compositions of peptides or proteins.[170] These procedures are advantageous in cases where samples contain amino acids easily decomposed by acid or alkali, such as tryptophan, and also for the detection of D-amino acids. By using a column packed with extracellular proteolytic enzyme from *Arthrobacter* immobilized on CNBr-activated Sepharose, limited hydrolysis of myoglobin was carried out.[171] Leucine aminopeptidase immobilized by physical adsorption on calcium phosphate gel was used to determine the amino acid sequence of low-molecular-weight peptides or for the hydrolysis of peptides.[172] In addition, trypsin immobilized on ethylene-maleic anhydride copolymer was used for the structural analysis of myosin and meromyosin.[173]

B. Nucleic acids

For the structural analysis of nucleic acids, immobilized phosphatase has been used.[174] Alkaline phosphatase immobilized by covalent binding with various polymers was proposed as a tool for the structural analysis of ribonucleic acid (RNA). Immobilized ribonuclease T_1 might also be useful for the structural analysis of RNA, as this enzyme hydrolyzes only the phosphodiester bond of guanosine.[175]

Further, phosphomonoesterase and phosphodiesterase immobilized on the triazine derivative of DEAE-cellulose were utilized for quantitative determination of the nucleotide composition and 5'-terminal analysis of oligonucleotides.[176]

4.3.3 Elucidation of the Mechanisms of Enzyme Reaction and Enzyme Function

As immobilized enzymes can be recovered easily from reaction mixtures, they can be conveniently used for the elucidation of complex enzyme reaction mechanisms.

A. Blood coagulation and liquefaction

One of the most effective applications of immobilized enzymes for the elucidation of complex enzyme reaction mechanisms has been the clarification of the mechanisms of blood coagulation and liquefaction.

Among the enzymes participating in blood coagulation and liquefaction, those which have been immobilized for investigations of the reaction mechanism are shown in Fig. 4.35. Fibrin is produced from fibrinogen by the action of thrombin, and blood coagulation is caused by a clot formed from fibrin via several subsequent reaction steps. Thrombin is formed from the precursor, prothrombin, by the action of an accelerator (factor V) which is produced by the activation of several factors in series starting from thrombokinase. This conversion of prothrombin to thrombin is also induced by trypsin.

On the other hand, coagulated blood is liquefied by the action of plasmin which is produced from plasminogen in the blood by urokinase, a kind of proteolytic enzyme. This hydrolytic reaction is caused by streptokinase from *Actinomycetes* as well as urokinase.

In order to investigate these complicated blood coagulation systems, immobilized trypsin was used, and the activation mechanisms of pro-

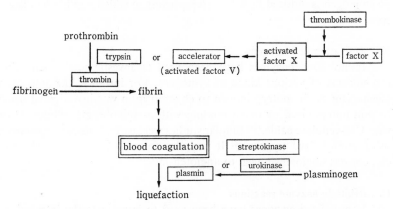

Fig. 4.35 Proteolytic enzymes participating in blood coagulation and liquefaction.
Square frames show enzymes which have so far been immobilized.

thrombin and factor X were investigated.[177,178] To clarify the activation mechanism of prothrombin, prothrombin was immobilized by diazo binding with amino acid copolymer to permit rapid removal of thrombin formed by the activation reaction, and the time course of activation was studied.[179] As the immobilized thrombin prepared by diazo binding with *m*-aminobenzoyloxymethyl cellulose had the ability to coagulate blood and also had esterase activity hydrolyzing tosyl-L-arginine methylester, it was useful for clarification of the role of thrombin in blood coagulation.[179] By using immobilized thrombokinase containing activation factor X prepared with CNBr-activated Sepharose, the mechanism of production of thrombin was investigated.[180]

On the other hand, as regards liquefaction, the mechanism of formation of plasmin, which plays a most important role in liquefaction, has been studied. For example, Rimon and his co-workers[181,182] prepared immobilized streptokinase by diazo binding with amino acid copolymer, and attempted to elucidate the complicated activation mechanism of blood plasminogen by using the immobilized enzyme. In order to study the difference in the activations of plasminogen to plasmin by streptokinase and urokinase, both enzymes were separately immobilized with CM-cellulose azide, and it was found that both immobilized enzymes showed the same activity toward plasminogen.[183] Urokinase immobilized on CNBr-activated Sepharose was also used to study the activation of plasminogen to plasmin.[184] Different degrees of activation of plasmin were obtained by passing plasminogen through the immobilized urokinase column at various flow rates, and it was found that the activation of plasminogen is induced by a two-step process in which peptide bonds are split.

B.   Activation of zymogens

Immobilized enzymes have been used to investigate the activation mechanisms of various kinds of zymogens. For instance, the activation mechanism of chymotrypsinogen to chymotrypsin was investigated using trypsin immobilized by diazo binding with *p*-aminobenzyl cellulose[59] or with CM-cellulose azide.[185] Similar studies were carried out for pepsinogen using immobilized trypsin.[186] It was also found that immobilized renin can convert angiotensinogen to angiotensin.[187]

C.   Multiple enzyme reactions

Immobilized enzymes have been used to investigate the interactions between enzymes in multi-enzyme reaction systems. For example, glucose oxidase was immobilized with trypsin or urease by the entrapping method using polyacrylamide gel, and changes in the pH-activity curve were studied.[188] It was found that the formation of protons in the presence of trypsin or urease caused the pH dependence of glucose oxidase activity to differ from that of the native enzyme or that of immobilized glucose oxidase alone.

Various kinetic analysis of multi-enzyme systems as models of mito-chondria have been carried out.[189] Malate dehydrogenase (Eq. 4.35) and citrate synthase (Eq. 4.36) were simultaneously immobilized on the same matrix by using CNBr–activated Sephadex or Sepharose, or by entrapping in polyacrylamide gel (overall reaction, Eq. 4.37), and in addition, these two enzymes and lactate dehydrogenase (Eq. 4.38) as an NAD-generating system were simultaneously immobilized by the same method. The reaction systems are shown in Eq. 4.39. Comparisons of the reaction efficiency by Eqs. 4.37 and 4.39 and of the reaction efficiency of immobilized systems having three kinds of enzyme on the same matrix with that of the native enzyme system were carried out. The immobilized enzyme system was

$$\text{malic acid} + NAD^+ \xrightarrow{\text{malate dehydrogenase}} \text{oxaloacetic acid} + NADH + H^+ \quad (4.35)$$

$$H_2O + \text{oxaloacetic acid} + \text{acetyl CoA} \xrightarrow{\text{citrate synthase}} \text{citric acid} + \text{reduced CoA} + H^+ \quad (4.36)$$

(Eqs. 4.35+4.36) $H_2O + \text{malic acid} + NAD^+ + \text{acetyl CoA}$
$\longrightarrow \text{citric acid} + \text{reduced CoA} + NADH + 2H^+$ (4.37)

$$H^+ + NADH + \text{pyruvic acid} \xrightarrow{\text{lactate dehydrogenase}} NAD^+ + \text{lactic acid} \quad (4.38)$$

(Eqs. 4.37+4.38) $H_2O + \text{malic acid} + \text{acetyl CoA} + \text{pyruvic acid}$
$\longrightarrow \text{citric acid} + \text{lactic acid} + \text{reduced CoA} + H^+$ (4.39)

found to be superior to the enzyme system in solution, and Eq. 4.39, including an NAD-regenerating system is more efficient than Eq. 4.37.

These results are presented schematically in Fig. 4.36. The figure shows that simultaneous immobilization of these enzymes on the same matrix is more efficient than separate immobilization. It is also more efficient than soluble enzymes.

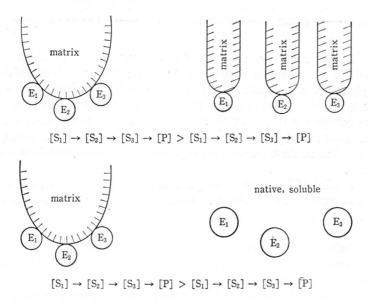

$$[S_1] \rightarrow [S_2] \rightarrow [S_3] \rightarrow [P] > [S_1] \rightarrow [S_2] \rightarrow [S_3] \rightarrow [P]$$

$$[S_1] \rightarrow [S_2] \rightarrow [S_3] \rightarrow [P] > [S_1] \rightarrow [S_2] \rightarrow [S_3] \rightarrow [P]$$

Fig. 4.36 Superiority of an immobilized multi-enzyme system in consecutive enzyme reactions.

D. Models of membrane-bound enzymes

In living organisms, not all enzymes exhibit catalytic activity in the soluble form; some exhibit their activity in a bound form to cellular particles or membranes, as is the case for respiratory enzymes, and enzymes participating in photosynthesis, protein synthesis and active transport. Thus, immobilized enzymes can be considered to be models for such bound enzymes.

Katchalski and his co-workers[190~192] prepared a papain membrane by adsorption of papain on a collodion membrane and subsequent cross-linking using bisdiazobenzidine; using this immobilized enzyme, they carried out kinetic studies on the relation between enzyme activity and pH, substrate concentration, and membrane permeability, in order to elucidate the mechanisms of enzyme action.

As a model system, enzyme membrane prepared by binding glutamate dehydrogenase to the azide derivative of collagen membrane,[193] or phosphorylase membrane prepared by using CNBr–activated Sephadex[194] has been used. In addition, three kinds of enzymes (Eq. 4.40) have been immobilized, and the rate of formation of gluconolactone-6-phosphate from lactose was determined.[195]

$$\underset{}{\text{lactose}} \xrightarrow{\text{lactase}} \text{glucose} \xrightarrow{\text{hexokinase}} \text{glucose-6-phosphate} \xrightarrow{\underset{\text{dehydrogenase}}{\text{glucose-6-phosphate}}} \text{gluconolactone}$$
(4.40)

It was found that the immobilized enzyme system in which three kinds of enzymes were simultaneously immobilized on CNBr–activated Sepharose was more efficient than that in which the enzymes were separately immobilized, then mixed. This suggests that in the case of consecutive enzymes participating in biosynthesis or metabolic conversion of biological materials in organisms, the most efficient catalytic activity is attained when these enzymes are all bound to the same particle or membrane.

Studies of immobilized enzymes as model systems for membrane-bound enzymes are expected to become increasingly important in biochemical research.

E. Subunits of enzymes

To investigate the functions of enzyme subunits, immobilized enzymes have been utilized.[194,196~199]

For example, Chan et al.[196,197] prepared immobilized aldolase consisting of 4 subunits by using CNBr–activated Sepharose, as shown in Fig. 4.37 (a), and studied the function of the subunits by treating the immobilized enzyme with urea. When the immobilized aldolase was treated with 8 M urea containing mercaptoethanol, 3 subunits not participating in the binding with the matrix were liberated from the remaining subunit, and the one immobilized subunit was denatured as shown in (b). By dialyzing this denatured immobilized subunit to remove urea, the single immobilized subunit (c) could be obtained. Aldolase was added to the denatured immobilized subunit suspended in 8 M urea and the mixture was then dialyzed to give reconstituted immobilized aldolase (d). The immobilized aldolase, immobilized subunit, and reconstructed immobilized aldolase showed specific activities of 4.5, 1.6 and 2.2 units per mg of protein, respectively.

Fukui et al.[198] utilized immobilized enzymes to investigate the dissociation and association of subunits of aspartate–β-decarboxylase from Pseudomonas dacunhae, and the activation mechanism of this enzyme by α-keto acids. This enzyme consists of 16 subunits, and reversibly disso-

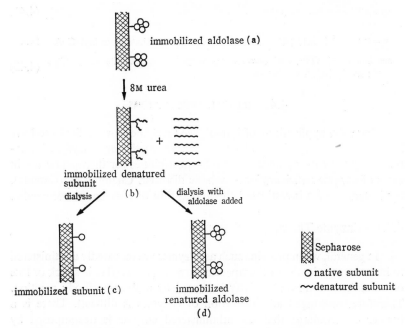

Fig. 4.37 Scheme for immobilized aldolase derivatives.[196]

ciates into two octomer molecules in the presence of 1 M guanidine hydrochloride. Both subunit preparations were immobilized on CNBr–activated Sepharose via a single subunit, and the properties of the resulting immobilized preparations were compared. It was found that the specific activity of the immobilized 8-subunit molecule was about 60% of that of the immobilized 16-subunit molecule, indicating an increase of specific activity on association of the subunits; the immobilized 8-subunit molecule was activated by α-keto acids to the same extent as the native enzyme. This indicates that the activation mechanism by α-keto acids is independent of the dissociation and association of subunits. This suggests a direction for future studies on the function of allosteric enzymes.

F. Succinyl–CoA synthetase

The reaction mechanism of succinyl–CoA synthetase is considered to proceed in two steps, as follows, though there is no direct evidence.[200] Hence, both the enzyme and ATP were immobilized by using CNBr–activated Sepharose in order to obtain direct evidence for the validity of Eq. 4.41. The formation of E–P was successfully demonstrated.

$$\text{\textcircled{E}} + ATP \overset{Mg^{2+}}{\rightleftharpoons} \text{\textcircled{E}} - P + ADP \tag{4.41}$$

$$\text{\textcircled{E}} - P + \text{succinic acid} + CoA \overset{Mg^{2+}}{\rightleftharpoons} \text{\textcircled{E}} + \text{inorganic phosphate} + \text{succinyl--CoA} \tag{4.42}$$

$$\text{succinic acid} + ATP + CoA \rightleftharpoons ADP + \text{inorganic phosphate} + \text{succinyl--CoA} \tag{4.43}$$
$$\text{\textcircled{E}}: \text{succinyl-CoA synthetase}$$

4.4 MEDICAL APPLICATIONS

Recently, applications of immobilized enzymes for medical use have been extensively studied and developed. In particular, enzymes immobilized by the entrapping method offer possibilities for therapeutic use in cases of enzyme deficiency or metabolic disorder. Chang *et al.* (Canada), for instance, have investigated the application of enzyme microcapsules.

4.4.1 Enzyme Therapy

In general, when proteins such as enzymes are repeatedly administered to humans or animals for therapeutic purposes, anaphylactic shock or loss of administered enzyme activity is caused by antigen-antibody reaction. Therefore, prolonged administration of enzymes is difficult. There is a further disadvantage that the administered enzyme is decomposed by various proteolytic enzymes and loses activity rather quickly. To overcome these problems, enzyme therapy by means of a blood extracorporeal shunt system using enzyme microcapsules or enzyme tubes has been investigated.

A. Enzyme microcapsules

The use of enzyme microcapsules, where the enzyme is entrapped inside a semipermeable polymer membrane, offers various advantages. That is, as shown in Fig. 4.38, a low-molecular-weight substrate can pass

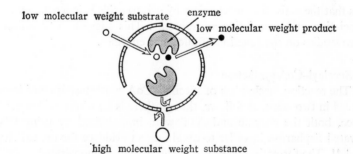

Fig. 4.38 Schematic diagram of the action of enzyme microcapsules.

freely through the capsule membrane, and after enzyme reaction in the capsule, the product passes out of the capsule, down the concentration gradient. The enzyme itself, being a polymer, cannot pass through the membrane, and is retained in the capsule. Similarly, external high-molecular-weight substances such as proteolytic enzymes, antibodies and antibody-producing cells cannot enter the capsule. Thus, in the case of enzyme microcapsules, an immunological reaction can be avoided, while the enzyme retains high activity toward a membrane-permeable, low-molecular-weight substrate for a long period. Examples of the application of enzyme microcapsules for therapy are as follows.

Chang and Poznansky[201-203] prepared catalase microcapsules by the phase separation method using collodion, and studied the effect of administration of the microcapsules to acatalasemic mice. Immediately after the intraperitoneal injection of catalase in soluble or microencapsulated form to acatalasemic mice, each mouse received a subcutaneous injection of sodium perborate, a substrate of catalase. After 20 min, each animal was homogenized, and the amount of sodium perborate retained in the body was measured. As shown in Fig. 4.39, remaining sodium perborate amounted to 7% for catalase in soluble form, and 16% for catalase in microencapsulated form. These values indicate very rapid removal with the control, which was 70%. Although repeated injection of catalase in soluble form led the death of the animal, in the case of catalase microcapsules, neither antibody formation nor side effects were observed.

Chang *et al.*[204,205] prepared asparaginase microcapsules, and at-

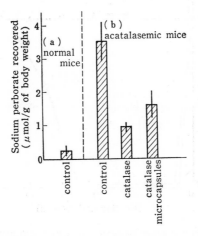

Fig. 4.39 Efficiency of catalase microcapsules in treating acatalasemic mice.[202]

tempted to use them for the therapy of leukemia. Namely, physiological saline, asparaginase in soluble form and asparaginase microcapsules prepared by using nylon membrane were each intraperitoneally injected into mice which had received a subcutaneous injection of 6C3HED lymphosarcoma, and the growth of the lymphosarcoma was followed. It was found that the implanted tumors first appeared after 9 days in the groups administered physiological saline, after 14 days in those administered asparaginase in soluble form, and after 69 days in the groups given asparaginase microcapsules. It is clear that asparaginase microcapsules effectively suppressed the growth of lymphosarcoma for a long period.

The changes of asparagine level in the blood after administration of asparaginase microcapsules were also studied.[206] That is, physiological saline, asparaginase in soluble form, asparaginase microencapsulated in cellulose nitrate, and the latter subsequently stabilized by glutaraldehyde treatment were intraperiotoneally administered to mice, and the time courses of asparagine level in the blood were observed. As shown in Fig. 4.40, asparaginase microcapsules were found to be effective for a long time; asparaginase in soluble form could maintain zero blood asparagine level for 2 days, while asparaginase microcapsules could do so for 6 days.

The authors[207] also prepared asparaginase microcapsules by the interfacial polymerization method using nylon or polyurea membranes, and investigated the applicability of these microcapsules for leukemia therapy. It was found that these asparaginase microcapsules are resistant

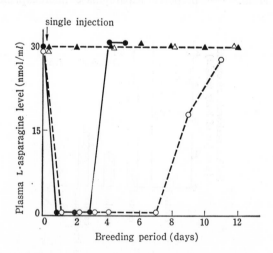

Fig. 4.40 Duration of action of asparaginase microcapsules.[206]
▲···▲, Physiological saline; ●—●, asparaginase solution; ○—○, asparaginase microcapsules; △—△, microcapsules not containing asparaginase.

to proteolytic enzymes, and microcapsules intraperitoneally injected into the rat shows no acute toxicity. The microcapsule membrane did not show antigenicity.

In some cases it is not desirable that enzyme microcapsules administered should remain for a long period in the recipient. Therefore, studies to develop membrane materials which can be decomposed and excreted from the body by some means at an appropriate time are required. As an approach to this, studies on the entrapment of asparaginase by red blood cell membranes have been reported.[208] In this case, disadvantages arising from the membrane were not observed.

Recently, investigations of liquid membrane microcapsules (liposomes) have been carried out, as described in Chapter 2; enzymes entrapped in such liposomes are being considered for therapeutic purposes.

B. Extracorporeal shunt systems

Although medical applications of immobilized enzymes by direct administration into the living body have been studied, utilization of immobilized enzymes for treatment of blood outside the body, i.e., extracorporeal shunt systems has been also attempted.

The authors[209] investigated the application of immobilized asparaginase prepared by entrapping in polyacrylamide gel, which is mechanically stronger than enzyme microcapsules, for the therapy of leukemia by means of an extracorporeal shunt system. This preparation was very stable toward proteolytic enzymes, and L-asparagine in the blood was completely decomposed when the blood was passed through a column packed with this immobilized enzyme. Further, coagulation of the blood was not observed during its passage through the column when an appropriate amount of heparin was added to the blood.

For the same purpose, an enzyme tube binding asparaginase was prepared.[210] The inner surface of a nylon tube (1 mm inside diameter and 2m in length) was partially hydrolyzed with hydrochloric acid, and after treatment with glutaraldehyde, asparaginase was covalently bound to this treated tube. When blood was passed through the asparaginase tube at 37°C at a flow rate of 2 ml/min, the plasma L-asparagine level was lowered to about 1/2. Complete decomposition of blood L-asparagine was assumed to be possible, if an asparaginase tube having higher activity could be prepared by increasing the amount of asparaginase bound to the tube. This asparaginase tube has very desirable properties as regards ease of blood flow, but the enzyme is directly in contact with the blood, and the possibility remains that decomposition by the action of proteolytic enzymes in the blood may occur.

In addition, an asparaginase plate was studied for the therapy of leukemia by means of an extracorporeal shunt system.[211] Namely, a polymethylmethacrylate plate ($20 \times 10 \times 4$ cm) was treated with γ-aminopropyltriethoxysilane and then activated with glutaraldehyde to give immobilized asparaginase. This asparaginase plate was set up as shown in Fig. 4.41. Extracorporeal shunt was carried out with a dog, and the time

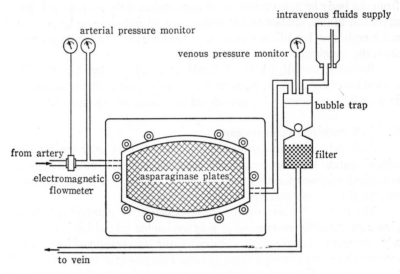

Fig. 4.41 Extracorporeal circuit with asparaginase plates.[211]

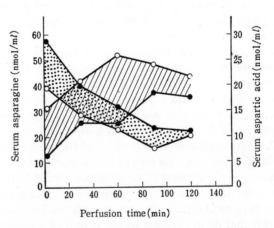

Fig. 4.42 Decomposition of blood asparagine by asparaginase plates in the case of a dog.[211] ●—●, Arterial levels; ○—○, venous levels; [▨], asparagine levels; [▨], aspartic acid levels.

courses of blood asparagine level and blood aspartic acid level were determined. The extracorporeal shunt was operated with addition of heparin for 2 to 3 h at a physiological rate of blood flow (150–250 ml/min). As shown in Fig. 4.42, when extracorporeal perfusion began, decreases of blood L-asparagine level in both veins and arteries occurred immediately, and the L-aspartic acid level in the blood increased with decreasing L-asparagine level. In this case, L-asparagine in the blood was not completely decomposed, though the reason for this was not determined. No toxic effects were observed during the perfusions, and histologic examination of the organs of animals subjected to perfusions showed no detectable abnormality. As in the case of the asparaginase tube, this method is also advantageous in offering low resistance to blood flow, but prolonged operation may be difficult due to the possibility of decomposition of asparaginase by proteolytic enzymes in the blood.

Asparaginase has also been immobilized on fibrin polymer, a biological material which may not stimulate the blood coagulation system or immunological system, for the therapy of leukemia.[212] Uricase was immobilized on aminoalkylated glass beads using glutaraldehyde, and this immobilized uricase was used for the therapy of hyperuricemia.[213]

As mentioned above, such an extracorporeal shunt system is superior to the direct administration of enzyme microcapsules as regards safety and technical factors, and is considered to have great potential. An extracorporeal shunt system may also be employed for emergency cases. The time required for therapy can be controlled easily, and specific detoxicant materials used for this system can be readily prepared, so that toxins such as drugs taken by accident can be effectively removed.

4.4.2 Artificial Organs

Recently, attempts to utilize immobilized enzymes for artificial organs have been made.

At present, the therapy of uremic disorders is carried out extracorporeally, removing metabolites such as urea and uric acid in the blood through a dialysis membrane, as shown in Fig. 4.43. However, this system is not very effective or economical because it requires cumbersome equipment capable of handling a large volume of dialysate.

On the other hand, Sparks et al.[214] investigated the removal of metabolites in the blood by the method shown in Fig. 4.44 in order to reduce the amount of dialysate and to minimize the instrumentation required. That is, urease and ion exchange resin are microencapsulated, and resulting microencapsulated urease and activated carbon are packed into a column as shown in Fig. 4.44. When this column and a compact dialyzer

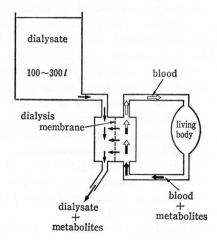

Fig. 4.43   Schematic diagram of a conventional artificial kidney.

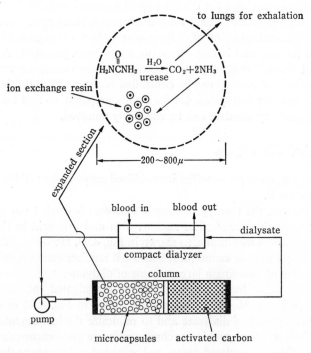

Fig. 4.44   Schematic diagram of a compact artificial kidney using urease microcapsules.[214]

are connected and dialysate is circulated through the column by means of a pump, urea in the dialysate is decomposed by the urease microcapsules to give ammonia and carbon dioxide and the ammonia is removed by the ion exchange resin. The carbon dioxide is exhaled via the lungs, so it is not necessary to remove it artificially. Metabolites in the dialysate other than urea are removed by the activated carbon. In this system, the volume of dialysate is markedly reduced compared with the conventional method, as the circulated dialysate can be reused.

Chang[215] carried out extracorporeal direct hemoperfusion with a dog, using a column packed with urease microcapsules in order to decompose urea in the blood. As shown in Fig. 4.45, when extracorporeal hemoperfusion was begun, ammonia concentration in the blood increased in parallel with the decrease of urea concentration in the blood, indicating the effectiveness of the urease microcapsules.

In these direct hemoperfusions, the problem of blood coagulation arises, so the preparation and use of microcapsules inducing no blood coagulation considerable interest.[216]

As mentioned above, urease microcapsules may be applicable for artificial kidneys, but cannot carry out all of the functions of natural kidneys, such as the removal of water or maintenance of electrolyte balance.

Further, it was reported that carbonic anhydrase and catalase immobilized on silicone or cellophane coated with a protein such as albumin or hemoglobin by means of glutaraldehyde treatment can be applied for artificial lungs.[217] These protein-coated carriers do not induce blood coagulation by contact with the blood, and permit efficient exchange of O_2

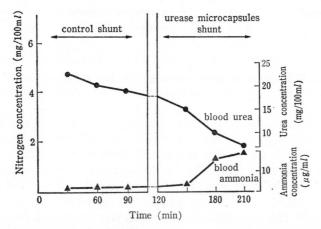

Fig. 4.45 Extracorporeal hemoperfusion using urease microcapsules.[215]

and CO_2 compared with the case of uncoated carriers.

In future, immobilized enzymes may be widely used for other artificial organs too, though the development of artificial organs exhibiting some specific function is likely, rather than organs having all of the complex functions of the natural organ.

As well as enzyme therapies and artificial organs, other applications of immobilized enzymes to medicine have been investigated. For example, urease and ethylene-maleic anhydride copolymer were separately micro-encapsulated in cellulose acetate butyrate, and oral administration of these microcapsules to dogs was performed for the therapy of uremic disorders.[218] Application of glucoamylase microcapsules prepared by liposomes consisting of lecithin, cholesterol and dicetyl phosphate was also reported for the therapy of glycogen storage disease.[219]

4.5   APPLICATIONS IN FOOD PROCESSING

Applications of enzymes in food manufacturing and processing have been increasing rapidly, and this is expected to continue.

4.5.1   Dairy Products

A.   Hydrolysis of lactose

Milk is a nutritionally excellent food, but lactose is contained at about 5% concentration and must be removed before feeding babies deficient in the enzyme hydrolyzing lactose (lactase or β-galactosidase) in the intestine. Dialysis of milk has been used, but the calorific value of the milk was reduced to 1/3, and various low-molecular-weight substances were also lost. Direct oral administration of this enzyme to lactase-deficient babies was also carried out, but this is not always satisfactory as allergic reaction can occur in some cases.

In order to resolve these problems, the use of immobilized lactase prepared by entrapping in polyacrylamide gel has been studied.[220] These studies have been carried forward especially in the U.S.A.,[221-223] Italy and Japan.

For instance, by using a column packed with lactase immobilized in polyacrylamide gel[224] or cyanuric chloride derivative,[225] continuous removal of lactose was investigated. Lactase from yeast has generally been used for treatment of lactase-deficient babies and also for immobilization.

In Italy, similar experiments were carried out using purified lactase from *Escherichia coli* or yeast immobilized by entrapping in triacetyl cellulose.[226] This immobilized enzyme was very stable, and retained the initial activity even after continuous operation for 80 days. It could thus be used for the industrial removal of lactose from milk. Utilization of immobilized lactase from *Aspergillus niger* for the continuous hydrolysis of lactose in acid whey has also been attempted.[227]

Lactase has been immobilized by various methods, as listed in Tables 2.1, 2.2, 2.4, 2.10, 2.11 and 2.16. The production of a sweet syrup consisting of glucose and galactose formed by the hydrolysis of lactose, and enhancement of the sweetness of milk used for ice cream or condensed milk by treatment with immobilized lactase have been investigated. Further, in order to enhance sweetness, whey was treated with immobilized lactase, and subsequently with immobilized glucose isomerase.

Since large amounts of lactose are produce, further applications of immobilized lactase are expected to be developed in future.

B. Production of cheese

The milk-clotting enzyme, rennin, which is used for cheese production, is in short supply, and microbial milk-clotting enzyme has also been used. In order to overcome this shortage of rennin, continuous production of milk curd by using immobilized rennin has been attempted. The clotting of milk by rennin involves two reactions, i.e., hydrolysis of casein, and subsequent association of the resulting micelles. If the first reaction can be performed by immobilized enzyme, separation of the reaction product becomes easy. Thus, rennin immobilized on CNBr-activated Sepharose or by carrier cross-linking using glutaraldehyde and AE-cellulose was prepared, and basic studies on the continuous clotting of milk were carried out.[228]

Some problems remain regarding the stability of the immobilized rennin and the type of enzyme reactor most suitable for the recovery of curd, and this process has not yet been industrialized.

C. Milk processing

Hydrogen peroxide is sometimes used for the sterilization of milk. In order to remove residual hydrogen peroxide after sterilization, a method utilizing catalase immobilized by covalent binding with carboxychloride resin was patented in the U.S.A.[229]

It is known that trypsin prevents the occurrence of an "oxidized" flavor. However, this enzyme has been of limited use, because of its high cost and the difficulty in inactivating the enzyme by pasteurization. To resolve these problems, utilization of immobilized enzyme prepared by covalent binding to porous glass was studied,[230] though the usefulness of this method has not yet been established.

4.5.2 Liquors

A. Hydrolysis of starch, proteins and polypeptides

Continuous pre-processing of alcoholic fermentation liquors using immobilized enzymes has been investigated. For instance, in the production of beer, attempts to hydrolyze starch by using immobilized amylases from microorganisms instead of malt, or to hydrolyze proteins or polypeptides by using immobilized proteolytic enzymes have been made. So far such processes are not used industrially.

B. Production of beer

Brewing of beer by using immobilized yeast has been investigated.[122] That is, on passing the wort at 15°C through a column packed with yeast immobilized on PVC and porous bricks, alcoholic fermentation occurred, and beer was produced. Further development of this kind of technique can be expected.

When beer is stored for a long period, it becomes turbid (so-called

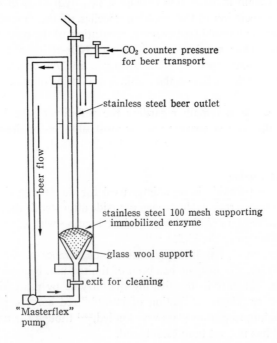

Fig. 4.46 Flow diagram for the continuous chill proofing of beer using immobilized papain.[231]

chill haze) due to the reaction of polyphenols and polypeptides contained in the beer. In order to prevent the formation of chill haze, papain is mainly used at present. However, when this proteolytic enzyme remains in beer for a long time, excess proteolysis occurs, and this is undesirable. Thus, for controlled treatment to prevent chill haze in beer, utilization of immobilized papain and immobilized polyphenol oxidase has been studied.

Fig. 4.46 shows the equipment devised by Witt *et al.*[231] for the removal of polypeptides in beer. That is, immobilized papain prepared by cross-linking with glutaraldehyde was packed into a glass column (15.25 × 104.14 cm), and about 23 l of beer was circulated through the column for several days at 0°C or 21°C to hydrolyze polypeptides. The resulting beer was stable for a long period, as in the case of ordinary papain treatment.

4.5.3   Miscellaneous Applications for Food Processing

To remove the bitterness of orange juice, naringinase is used. Naringinase is produced by microorganisms, and it decomposes naringine, a bitter substance in oranges, to rhamnose and purunine. Naringinase was immobilized on ethylene-maleic anhydride copolymer, and its application for this purpose was studied. The removal of bitterness was efficiently carried out by this method in both batch and column systems.[232]

A West German patent[233] describes how glucose oxidase and peroxidase bound to cellulose membrane were used for the removal of glucose in egg white albumin. This patent claimed that when glucose oxidase and catalase immobilized in a starch matrix are coated on the inner surface of a can, decomposition of mayonnaise in the can is retarded; no decomposition or changes in flavor could be observed after storage for 5 months. Further, tannase immobilized on aminoalkylated glass beads by means of glutaraldehyde has been used for the treatment of tea cream.[234] Other interesting reports include application of immobilized enzymes for the purification of sugar.[235] In this case, dextran existing in the juice from pressed cane causes various problems during the purification process. Accordingly, preliminary removal of dextran is desirable. Hence, a bacterial dextran-hydrolyzing enzyme was immobilized by using CM-cellulose azide or CNBr-activated cellulose, and the hydrolysis of dextran could be efficiently carried out by using this immobilized enzyme. In addition, α-galactosidase was adsorbed on nylon pellets treated with formic acid, and then treated with dimethyladipimide and methylacetimidate. The resulting immobilized preparation was used for the reduction of raffinose in beet sugar molasses.[236]

4.6   AFFINITY CHROMATOGRAPHY

Many biochemical studies involve elucidation of the structure and function of biologically active substances such as enzymes, nucleic acids, antibodies, hormones, and so forth. In these studies, isolation and purification of biologically active substances are often a necessary starting point. Separation procedures such as salting out, precipitation with organic solvents, electrophoresis, ion exchange chromatography, gel filtration chromatography, etc., are widely used for the isolation and purification of biologically active substances. However, all these methods utilize the physical and chemical characteristics of biologically active substances, such as solubility, molecular size, molecular shape and electrical properties, so the specificity of separation is low, and it is necessary to combine various fractionation procedures for a particular case.

On the other hand, "affinity chromatography" is a purification procedure based on the specific properties of biological substances, i.e., their biological specificity (*see* Table 4.14). Thus, the highly specific interactions shown in the table can be utilized for the separation and purification of biological substances by so-called "affintiy chromatography". As conventional chromatography utilizes the physical affinities between the carrier and target substance, this new method should perhaps be called "bio-affinity chromatography" or "biospecific affinity chromatography".

Although the technique of affinity chromatography has been used for the purification of antibodies in the field of immunochemistry for some time, its use did not spread widely. The reason for this is considered to be the difficulty of immobilization of one substance having affinity for another specific substance. Also, suitable carriers for immobilization were not easy to find.

Development of the immobilization techniques for enzymes described in Chapter 2, and of various hydrophilic carriers for immobilization devel-

TABLE 4.14   Substances having biological interactions

Substance showing interaction	Substance showing interaction
Enzyme	substrate
	reaction product
	competitive inhibitor
	coenzyme
	allosteric effector
Antigen	antibody
Hormone	receptor protein
Nucleic acid (DNA)	nucleic acid (*m*RNA)

oped by Porath and his co-workers in Sweden resolved many of these problems. Anfinsen and Cuatrecasas and their co-workers in the U.S.A. developed the technique of affinity chromatography by utilizing these procedures for the separation and purification of biological substances.

Descriptions of affinity chromatography have been already published by these investigators,[237~242] and in Japan the authors have published a book on affinity chromatography.[243,244]

In this section, affinity chromatography involving immobilized enzymes is mainly described.

4.6.1 Principles and Methods

A. Principles

The principles of affinity chromatography are shown schematically in Fig. 4.47. That is, a substance biologically related to the substance to be purified is selected as a ligand, then immobilized and used as the adsorbent

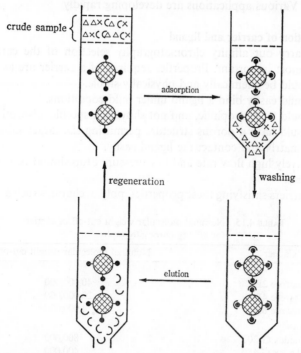

Fig. 4.47 Schematic diagram of affinity chromatography. ( ), Target substance; $\triangle_\triangle\times$, impurities; crosshatching, water-insoluble carrier; ●-●-, ●-◐-, ligand.

for affinity chromatography. The sample containing the substance to be purified is passed through a column packed with the adsorbent, as in the case of general chromatography. Only the substance having specific affinity toward the immobilized ligand is adsorbed on the water-insoluble carrier, as shown in the figure, while substances having no affinity flow through without any interaction. After removal of impurities by washing the adsorbent, the adsorbed substance can be eluted from the column by changing the ionic strength and pH or passing a solution containing substrate, inhibitor or reaction product through the column. Such a column can be easily regenerated by elution and washing, and reused for chromatography. By this method, a preparation of high purity can be obtained from a crude preparation in high yield by a single operation of affinity chromatography in some cases. Because of the high specificity of this method, it is possible in many cases to remove impurities which cannot be removed by conventional methods. Accordingly, for the isolation and purification of biological substances, affinity chromatography is one of the most promising methods available. Various applications are developing rapidly.

B. Selection of carrier and ligand

To carry out affinity chromatography selection of the carrier and ligand is most important. Properties required of a carrier are as follows.
1) It should be chemically and physically stable.
2) It should easily bind a ligand under mild conditions.
3) It should be hydrophilic, and not show non-specific adsorption.
4) It should have a porous structure permitting the target substance to enter the matrix and contact the ligand easily.
5) Relatively high flow rate and low pressure drop should be obtainable in a column.

As carriers satisfying these properties, polysaccharides such as agarose

TABLE 4.15 Nominal molecular weight cut-off of carriers for affinity chromatography

Carrier	Nominal molecular weight cut-off[†]
AGAROSE	
Sepharose 2B	40,000,000
Sepharose 4B	20,000,000
Sepharose 6B	4,000,000
DEXTRAN	
Sephadex G-200	800,000
Sephadex G-150	400,000
POLYACRYLAMIDE	
Bio-Gel P-300	500,000

† Molecular weight of a control globular protein.

and dextran, or polyacrylamide can be used. These carriers are commercially available for gel filtration of high-molecular-weight substances under the trade names Sepharose, Sephadex, and Bio-Gel, respectively. As the porosity of the gel network of these carriers can be estimated from the molecular weight cut-off (Table 4.15), selection of a carrier suitable for a particular ligand or adsorbing substance is possible. When an enzyme is selected as a ligand and immobilized by suitable methods, the resulting immobilized enzyme can be used for the isolation of substrates, competitive inhibitors, coenzymes, allosteric effectors, or antibodies of the enzyme.

C. Preparation of adsorbent

The immobilization of a ligand is carried out by the carrier-binding method in many cases; the immobilization of enzymes is described in Chapter 2. Immobilization of a ligand on a carrier should be carried out via a reactive group in the ligand molecule which does not participate in affinity binding with the substance to be purified, as in the case of immobilization of enzymes. Polysaccharide and polyacrylamide are often used as carriers, and the entrapping and cross-linking methods are often employed, as well as the carrier-binding method.

a. Carrier-binding Method
i) *Examples utilizing polysaccharides as carriers*

Equations 4.44 through 4.50 show reactions which have been carried out using polysaccharides as carriers for ligand immobilization. Ligands are shown enclosed in boxes ($\square$).

Among these procedures, Eq. 4.44 shows immobilization based on the azo binding of a ligand having a phenol group or imidazole group with a water-insoluble diazonium compound. The immobilization method shown in Eq. 4.45 was developed by Axén *et al.*[245] as described in Chapter 2. This is a widely used method based on binding between CNBr-activated polysaccharides and a ligand containing an amino group under mild conditions. This activated Sepharose is sold by Pharmacia Fine Chemicals. Equation 4.46 shows immobilization based on the alkylation of liganos having amino, phenol, and imidazole groups. Equations 4.47 and 4.48 are used for the immobilization of ligands having a carboxyl or amino group by using a condensing reagent such as a water-soluble carbodiimide.

Further, when chain structure consisting of an appropriate number of methylene residues is interposed between the carrier and ligand, this so-called spacer or arm is expected to enhance the overall affectiveness. That is, when a ligand is directly immobilized on a high-molecular-weight substance such as a polysaccharide, it becomes difficult for another high-molecular-weight compound such as an enzyme to approach close to the

$$(4.44)$$

$$(4.45)$$

$$(4.46)$$

$$\text{polysaccharide} \begin{cases} O \\ O \end{cases} C=NH \xrightarrow{NH_2(CH_2)_nNH_2} \text{polysaccharide} \begin{cases} -OCONH(CH_2)_nNH_2 \\ -OH \end{cases} \xrightarrow[\text{water-soluble carbodiimide}]{\boxed{R\text{-}COOH}}$$

$$\text{polysaccharide} \begin{cases} -OCONH(CH_2)_nNHCO-\boxed{R} \\ -OH \end{cases}$$

$$(4.47)$$

$$\text{polysaccharide} \begin{cases} -OCONH(CH_2)_nNH_2 \\ -OH \end{cases} \xrightarrow{O\diagdown\diagup O} \text{polysaccharide} \begin{cases} -OCONH(CH_2)_nNHCOCH_2CH_2COOH \\ -OH \end{cases}$$

$$\xrightarrow[\text{water-soluble carbodiimide}]{\boxed{R\text{-}NH_2}} \text{polysaccharide} \begin{cases} -OCONH(CH_2)_nNHCOCH_2CH_2CONH-\boxed{R} \\ -OH \end{cases}$$

$$(4.48)$$

ligand. In such a case, the affinity increases on interposing some chain structure (spacer) between the carrier and ligand, as shown in Fig. 4.48.[246]

Equation 4.49 shows a method suitable for the immobilization of ligands having an active halogen atom, and Eq. 4.50 shows the immobilization of ligands having a carboxyl group, as in the case of Eq. 4.47.

ii) Examples utilizing polyacrylamide as a carrier

Polyacrylamide and its derivative are also often used as carriers for the immobilization of ligands. Preparation procedures for derivatives suitable for the immobilization of ligands are presented in Eqs. 4.51 through 4.54.

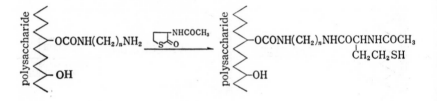

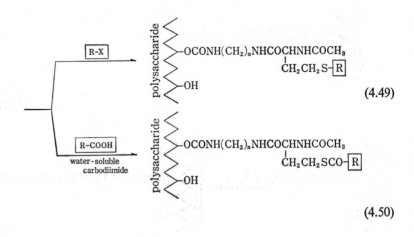

(4.49)

(4.50)

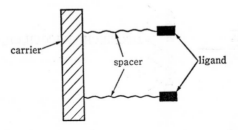

Fig. 4.48 Adsorbent prepared by interposing a spacer between a carrier and ligand.

$$\underset{\text{polyacrylamide}}{}-CONH_2 \xrightarrow[90^\circ C]{H_2N(CH_2)_nNH_2} \underset{\text{polyacrylamide}}{}-CONH(CH_2)_nNH_2$$

(4.51)

$$\text{polyacrylamide}-\text{CONHNH}_2 \xrightarrow[\text{pH 4}]{} \text{polyacrylamide}-\text{CONHNHCOCH}_2\text{CH}_2\text{COOH} \tag{4.52}$$

$$\text{polyacrylamide}-\text{CONH}_2 \xrightarrow[\text{60℃}]{\text{NaOH}} \text{polyacrylamide}-\text{COOH} \tag{4.53}$$

$$\text{polyacrylamide}-\text{CONH}_2 \xrightarrow[\text{50℃}]{\text{NH}_2\text{NH}_2} \text{polyacrylamide}-\text{CONHNH}_2 \xrightarrow{\text{HNO}_2} \text{polyacrylamide}-\text{CON}_3 \xrightarrow{\boxed{\text{R-NH}_2}} \text{polyacrylamide}-\text{CONH}\boxed{\text{R}} \tag{4.54}$$

Equation 4.51 shows the procedure for carriers having amino groups, and Eqs. 4.52 and 4.53 show procedures for carriers having carboxyl groups. Once these carriers are prepared, ligands having carboxyl or amino groups can be bound using a condensing reagent such as water-soluble carbodiimide, yielding adsorbents for affinity chromatography. Equation 4.54 shows the immobilization of ligands having an amino group by the peptide binding method.

b. Cross-linking and entrapping methods

The techniques of the cross-linking method using bifunctional reagents and the entrapping method using polyacrylamide gel or starch as described in Chapter 2 have also been utilized. In the cross-linking method, the enzyme is mainly used as a ligand, so immobilized enzyme itself is the adsorbent in this case. In the entrapping method, the ligand must be a high-molecular-weight compound which cannot pass through the gel lattice or semipermeable membrane, while the substance adsorbed must be a low-molecular-weight compound which can permeate the gel lattice or membrane.

D.   Adsorption and elution

The adsorption of a target compound on an adsorbent for affinity chromatography is greatly affected by pH, ionic strength, temperature, and so forth. Therefore, preliminary experiments on a small scale are necessary to determine the optimum conditions. As the purity of the substance to be purified also affects the adsorption, the removal of impurities by chemical or physical means can assist the highly specific adsorption of the substance to be purified in many cases.

Elution of the adsorbed substance from the adsorbent column can be carried out by reversing the conditions of adsorption, i.e., reducing the adsorption force between the adsorbent and adsorbed substance. Generally, elution is carried out by changing the ionic strength, pH, temperature or some combination of these factors. Further, in the presence of protein-denaturing agents such as urea, guanidine hydrochloride, etc., elution can be performed more smoothly in some cases. If the substance to be eluted is a protein, such as an enzyme, these denaturing agents must be removed immediately after elution to prevent denaturation. When the adsorbed substance is an enzyme, specific elution can be carried out using a substrate, reaction product, competitive inhibitor, coenzyme, or allosteric effector of the enzyme. As this elution method shows higher specificity than elution by alteration of the ionic strength, pH or temperature, it yields a product of higher purity. Various amino acids, nucleic acid bases and their derivatives, and coenzymes immobilized on polysaccharide are now commercially available as adsorbents for affinity chromatography.

4.6.2   Isolation of Enzyme Inhibitors

The isolation and purification of enzyme inhibitors by affinity chromatography using immobilized enzymes are interesting as an application of immobilized enzymes. This was first attempted by Werle and his co-workers.[247-249] Proteolytic enzymes such as trypsin, α-chymotrypsin, and kallikrein were immobilized by the peptide binding method using ethylene-maleic anhydride copolymer. When impure solution containing inhibitor was passed through a column packed with immobilized proteolytic enzyme, only specific inhibitor showing affinity for the enzyme was adsorbed. By this method, almost pure inhibitor was obtained in 90–95% yield, as shown in Table 4.16. Such a column could be reused repeatedly without loss of adsorption activity.

In the case of adsorption of a protein-like trypsin inhibitor by using immobilized trypsin, there is the disadvantage that the active center of the inhibitor can easily be hydrolyzed by trypsin. However, it is well known that some kinds of trypsin inhibitor can bind not only with trypsin but

TABLE 4.16 Isolation of inhibitors by using immobilized
proteolytic enzymes[249]

Immobilized enzyme	Isolated inhibitor and its source	Yield of inhibitor (%)
Trypsin	TRYPSIN INHIBITOR	
	human pancreas	95
	porcine pancrease	90–95
	dog pancreas	90
	bovine pancrease	90–95
	mouse sperma	75–85
Chymotrypsin	TRYPSIN-KALLIKREIN INHIBITOR	
	bovine pancreas	90
Kallikirein	TRYPSIN-KALLIKREIN INHIBITOR	
	bovine organs	60–90

also with trypsinogen, the precursor of trypsin. By using immobilized trypsinogen, purification of protein-like trypsin inhibitors can be carried out.[250] That is, the adsorbent for affinity chromatography was prepared by immobilizing trypsinogen on CNB-activated Sepharose. By using this adsorbent, purification of the trypsin inhibitor in peanuts was performed more easily than with immobilized trypsin.

Purification of an inhibitor of deoxyribonuclease I existing in large amounts in calf thymus has been carried out by using a column packed with deoxyribonuclease I immobilized on CNBr-activated Sepharose.[251] Further, immobilized elastase prepared with CNBr-activated Sepharose was used for the isolation of the elastase inhibitor in potato.[252]

4.6.3 Isolation of Enzymes

Most papers on affinity chromatography deal with the isolation of enzymes. Suitable adsorbents for the isolation of enzymes include a substrate, reaction product, inhibitor, allosteric effector, coenzyme, antibody, and so on. Typical examples are as follows.

A. Examples utilizing a substrate or reaction product as a ligand

The relation between an enzyme and its substrate is highly specific and has been likened to the relation between a key and keyhole. The purification of enzymes by affinity chromatography using immobilized substrate is therefore advantageous.

The phenomenon of adsorption of enzymes on a natural water-insoluble substrate or immobilized substrate has been known for many years. That is, in 1910, Starkenstein[253] observed that amylase was adsorbed on water-insoluble starch, and carried out the purification of amylase by using water-insoluble starch, which is a naturally insoluble substrate. In 1949, amylase adsorbed on insoluble starch was found to be eluted by soluble starch.[254] The first attempt to purify an enzyme by affinity chromatography was made by Lerman[255] in 1953. That is, purification of tyrosinase was carried out by using p-azophenol cellulose as an adsorbent. In this case, the phenol group in the adsorbent is considered to act as a substrate analog of tyrosinase.

DNA polymerase synthesizes long-chain DNA by polymerization of nucleotides utilizing previously synthesized DNA as a template. Immobilized DNA was used for the purification of this enzyme: DNA of Hela cells was treated with exonuclease and the resulting DNA preparation was immobilized on Sepharose. By using this adsorbent, purification of DNA polymerase of Hela cells was achieved.[256] Further, DNA of *Escherichia coli* was entrapped in the network structure of polyacrylamide gel, and used for the purification of *E. coli* DNA polymerase. The specific activity of the enzyme was increased 200 times by a single chromatography.[257]

The proteolytic enzyme contained in wheat flour in small amounts reacts with other proteins and induces coagulation, so its purification is difficult by conventional procedures. However, hemoglobin, a substrate of this enzyme, was immobilized on Sepharose and used for the purification of the enzyme.[258] In the case of purification of an enzyme using an adsorbent with the substrate as a ligand, the substrate may be degraded by the enzyme. Therefore, in the case of purification of the wheat flour proteolytic enzyme, adsorption of enzyme was carried out at pH 5.5, away from the optimum pH (3.8) for enzyme reaction, in order to prevent degradation of the ligand.

The authors also carried out the isolation and purification of enzymes by using immobilized substrate and reaction product, and analyzed the characteristics and group specificity of affinity chromatography.[259] For the purification of enzymes relating to the metabolism of L-aspartic acid, shown in Fig. 4.49, N-(ω-aminohexyl)-L-aspartic acid (7) was synthesized as a ligand, with hexamethylene acting as a spacer. This aspartate derivative was immobilized on CNBr-activated Sepharose, and used as an adsorbent for affinity chromatography. As shown in Fig. 4.49, affinity chromatographies were carried out with this adsorbent for aspartase,

$$H_2N(CH_2)_6NHCHCH_2COOH$$
$$\overset{|}{COOH}$$
$$7$$

aspartate-4-decarboxylase, 2 kinds of asparaginase from different sources, and fumarase, which is not related to the metabolism of L-aspartic acid.

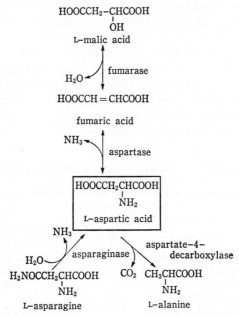

Fig. 4.49 Enzymes related to the metabolism of L-aspartic acid.

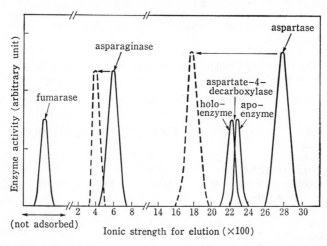

Fig. 4.50 Affinity chromatography of various enzymes using an L-aspartic acid-Sepharose column.[259]

Each enzyme dissolved in sodium acetate buffer (pH 7.0; ionic strength, 0.01) was applied to the column equilibrated with the above buffer.

Solid line: elution pattern with buffer containing sodium chloride.

Dotted line: elution pattern with buffer containing substrate (fumaric acid for aspartase) or reaction product (L-aspartic acid for asparaginase)

The results are summarized in Fig. 4.50. It is clear that aspartase, aspartate-4-decarboxylase, and asparaginase are well adsorbed on this adsorbent, but adsorption of fumarase, which is not related to the metabolism of L-aspartic acid, does not occur. Ionic strengths at which adsorbed enzymes are eluted are compared in Table 4.17. These ionic strengths seem to correspond to apparent affinities between the adsorbent and enzyme. Fig. 4.50 shows the elution of asparaginase by L-aspartic acid and the elution of aspartase by fumaric acid. It is clear that these enzymes are easily eluted by the specific product and substrate compared with elution by sodium

TABLE 4.17   Affinities of various enzymes toward L-aspartic acid-Sepharose[259]

Enzyme	Affinity Ionic strength at which enzyme is eluted by NaCl
Fumarase	not adsorbed
Proteus vulgaris asparaginase	0.060
Escherichia coli asparaginase	0.062
Modified asparaginase[†]	>0.210
Aspartate-4-decarboxylase	
(Holoenzyme)	0.228
(Apoenzyme)	0.233
Aspartase	>0.275

† Tyrosine residues of asparaginase from *P. vulgaris* were modified with tetranitromethane.

TABLE 4.18   Purification of L-asparaginase from *Proteus vulgaris*[259]

Purification step	Volume (ml)	Total protein (mg)	Specific activity (unit[†]/mg)	Total activity (unit[†])	Yield (%)
1. Cell-free extract	1,050	19,820	1.1	21,800	100
2. Supernatant after treatment at pH 4.5	930	9,420	1.9	17,900	82
3. Ammonium sulfate fractionation (50–90% saturation)	42	1,020	13.2	13,500	62
4. Ethanol fractionation (33–50% concentration)	50	149	70	10,460	48
5. L-Aspartic acid–Sepharose column	24	30	310	9,160	42
6. Crystals	1	27	300	8,080	37

† One enzyme unit is defined as the amount of enzyme which produces one μmol of ammonia per min at 37°C, at pH 8.4.

chloride. This indicates that adsorption of these enzymes does not depend on electrostatic and ionic binding, but is due to so-called specific bio-affinity. Aspberg and Porath[260] first reported the adsorption of various glycoproteins on an adsorbent having concanavalin A as the ligand. Subsequently, Uren,[261] and Lowe and Dean[262] reported the adsorption of various proteolytic enzymes on an adsorbent having glycyl-D-phenylalanine as the ligand, and the adsorption of various dehydrogenases on an adsorbent having NAD as the ligand, respectively. However, group specificity in affinity chromatography of enzymes having different enzyme actions was first clarified by the authors, as described here, for the enzymes related to aspartic acid metabolism. Also, as shown in Table 4.18, pH treatment, salting out and fractionation with organic solvent were combined with affinity chromatography using the above L-aspartic acid-type adsorbent, and industrial purification was carried out of the asparaginase from *Proteus vulgaris*, which has anti-leukemic action and is immunologically different from *Escherichia coli* asparaginase.

B. Examples utilizing an enzyme inhibitor as a ligand

For the purification of enzymes by affinity chromatography, many workers have used adsorbents having enzyme inhibitor as a ligand.

Cuatrecasas *et al.*[263] obtained very pure carboxypeptidase A in high yield by affinity chromatography using a column packed with the inhibitor of carboxypeptidase A immobilized by covalent binding to CNBr-activated Sepharose. As shown in Fig. 4.51, this enzyme was not adsorbed on Sepharose, but was specifically adsorbed on adsorbent having the inhibitor as a ligand. Carboxypeptidase B has very similar physical properties to carboxypeptidase A, and separation of the two enzymes is difficult by usual purification methods. However, carboxypeptidase B was not adsorbed on this adsorbent, as shown in Fig. 4.51 (c), so separation of the two enzymes can be easily carried out by using this adsorbent, which clearly has high specificity.

Cuatrecasas *et al.*[263] prepared an adsorbent (**8**) by binding D-tryptophan methylester, an inhibitor of α-chymotrypsin, to CNBr-activated Sepharose. Affinity chromatography was carried out using this adsorbent, but satisfactory results were not obtained. However, α-chymotrypsin was specifically adsorbed on an adsorbent having ε-aminocaproyl-D-tryptophan methyl ester (**9**) as the ligand. Therefore, the ε-aminocaproyl moiety is considered to act as a spacer.

The purification of nucleic acid-related enzymes by using an adsorbent prepared by binding 3′-(4-amino-phenylphosphoryl)-deoxythymidine-5′-phosphate (**10**) to CNBr-activated Sepharose was attempted by Cuatrecasas *et al.*[263] This adsorbent shows strong affinity (K_i value, 10^{-6} M)

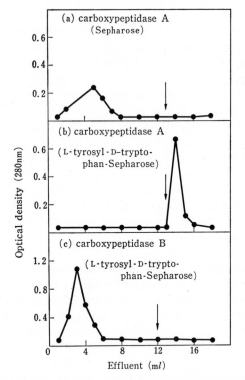

Fig. 4.51 Affinity chromatography of carboxypeptidase A.[263]
Elution was accomplished with 0.1 M acetic acid (arrow).

Sepharose—OCONHCHCOOCH$_3$
                    |
                   CH$_2$—indole(N-H)

8

Sepharose—OCONH(CH$_2$)$_5$CONHCHCOOCH$_3$
                                |
                               CH$_2$—indole(N-H)

9

toward an extracellular nuclease from pathogenic *Staphylococcus,* and is useful for the purification of this enzyme.

As well as adsorbents prepared by using synthetic inhibitors, adsorbents prepared by using natural enzyme inhibitors have been used for the purification of enzymes. For example, ovomucoid, a kind of trypsin inhibitor contained in egg white, was bound to CNBr-activated Sepharose, and the resulting adsorbent was used for affinity chromatography.[264] As

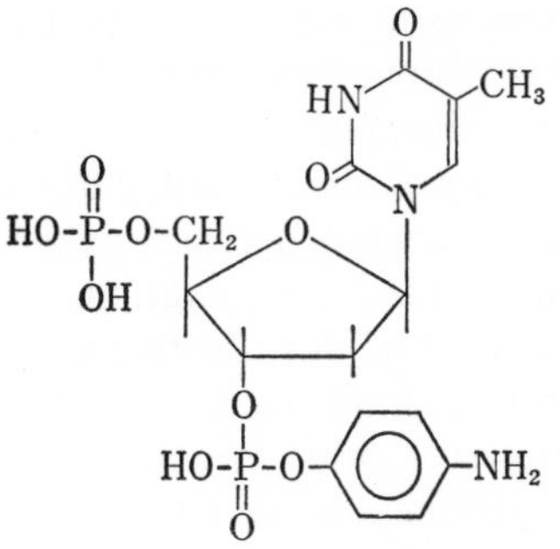

10

the ovomucoid binds with trypsin at a molar ratio of 1:1 and inactivates the enzyme, this adsorbent can be effectively used for the removal of very small amounts of chymotrypsin contaminating trypsin.

In order to elucidate the relation between enzyme and inhibitor, Claeyssens *et al.*[265] carried out an interesting experiment. That is, *p*-aminobenzyl–1-thio-β-D-xylopyranoside (**11**), a competitive inhibitor of β-D-xylosidase, was immobilized on CNBr-activated Sepharose, and the resulting adsorbent was used for the affinity chromatography of this enzyme. However, in the case of immobilized *p*-aminophenyl–1-thio-β-D-xylopyranoside (**12**), lacking one methylene residue compared with the previous compound (**11**), the enzyme was not completely adsorbed.

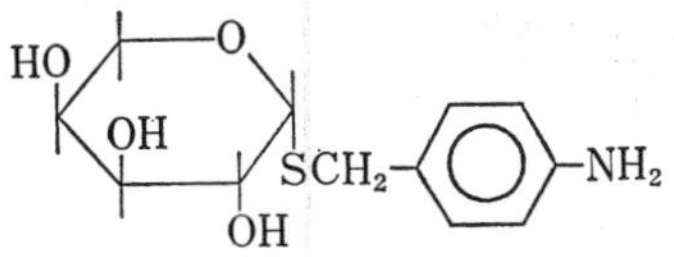

11 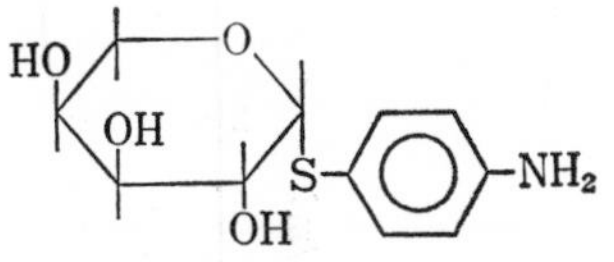   **12**

C.   Examples utilizing an allosteric effector as a ligand

As well as an active center responsible for the enzyme activity, allosteric enzymes have an allosteric or regulatory site participating in binding with a specific metabolite other than the substrate. A compound binding with the allosteric site is called an "allosteric effector", and binding of this effector to the allosteric site of the enzyme changes the reactivity of its active center, causing inhibition or activation of the enzyme. This mechanism operates to control metabolism *in vivo*. Accordingly, the purification of allosteric enzymes may be possible by using an immobilized allosteric effector of the enzyme.

3-Deoxy-D-arabino-hepturonate-7-phosphate synthetase is an allosteric enzyme. This enzyme participates in the regulation of aromatic amino acid biosynthesis, and is inhibited by the final reaction products of this metabolic system, i.e., tyrosine, phenylalanine, tryptophan, etc. (so-called feedback inhibition). Affinity chromatography of this enzyme was carried out by using an adsorbent having L-tyrosine as the ligand. It was found two components exist in this enzyme preparation, of which one is adsorbed and the other is not adsorbed on this adsorbent.[266] The enzyme preparation before affinity chromatography was equally inhibited by L-tyrosine and L-phenylalanine. However, after chromatography the adsorbed portion was inhibited by L-tyrosine, but not by L-phenylalanine. On the other hand, the unadsorbed portion was not inhibited by L-tyrosine but was by L-phenylalanine. That is, two kinds of isozymes were present in this enzyme preparation. Further, L-tryptophan, an allosteric effector of chorismate mutase, was immobilized as a ligand, and purification of this enzyme could be accomplished efficiently by using this adsorbent.[267]

D.   Examples utilizing a coenzyme as a ligand

The purification of enzymes having coenzymes has also been carried out by using the immobilized coenzyme.

For instance, NAD was bound to CNBr-activated Sepharose, and

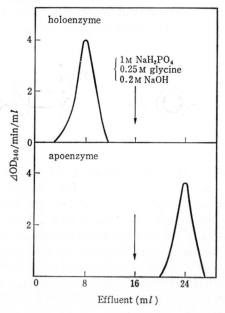

Fig. 4.52   Affinity chromatography of glutamate-oxaloacetate transaminase using N-ω-aminoalkyl-PMP-Sepharose.[269]

affinity chromatography of various dehydrogenases was carried out with the resulting adsorbent.[262] Glucose-6-phosphate dehydrogenase, lactate dehydrogenase, and L-threonine 3-dehydrogenase were purified from a crude extract of *Pseudomonas oxalaticus* by this method. L-Threonine 3-dehydrogenase was fractionated into two parts with this immobilized coenzyme, and the specific activities of the fractions were increased 150 times and 450 times.

The authors[268] prepared an adsorbent by linking coenzyme A as a ligand to CNBr–activated Sepharose, and purification of transacetylase of *Clostridium klyverii* was shown to be possible using this adsorbent. An adsorbent prepared by immobilizing pyridoxamine phosphate as a ligand on CNBr-activated Sepharose was also used for the purification of gluta-mate-oxaloacetate transaminase, which has pyridoxal phosphate as a coenzyme.[269] From the chromatographic pattern (Fig. 4.52), it is clear that the holoenzyme (containing coenzyme) is not adsorbed on this ad-sorbent, but the apoenzyme is adsorbed.

E. Examples utilizing an antibody as a ligand

When a protein such as an enzyme is administered to an animal, an antibody reacting specifically with the protein is formed by the animal. The resulting antibody specifically binds with the protein antigen. Thus, affinity chromatography using immobilized antibody may be possible. For example, anti-asparaginase protein of rabbit was immobilized on dia-zotized silanized glass, and purification of asparaginase from *Escherichia coli* was carried out using this adsorbent.[270]

4.6.4   Isolation of Antibodies

Before the concept of affinity chromatography had been established, the purification of antibodies was carried out using immobilized antigen as an immunoadsorbent. However, the isolation of the antibody of an enzyme using immobilized enzyme has been attempted recently.

For instance, the nuclease from *Staphylococcus aureus* and fragment P_3 consisting of amino acid residues from No. 49 to No. 149 obtained by limited hydrolysis of the nuclease with trypsin were bound to CNBr-activated Sepharose, and immobilized nuclease and immobilized P_3 were obtained. Rabbit serum containing anti-nuclease was passed through the immobilized nuclease column, and the column was thoroughly washed. When adsorbed antibody was eluted by alteration of the pH, 36% of the protein charged on the column was recovered. In a similar experiment using an immobilized P_3 column, 4–5% of the protein was recovered. This suggests that the fragment P_3 may contain the antigen-determining group.

Further, the P_3-binding capacity of the resulting anti-P_3 antibody was 6–10 times higher than that of anti-nuclease antibody.[57,271]

4.6.5 Isolation of Substrate

Immobilized enzymes have also been used for the isolation of substrates. For instance, pure histidyl-*t*RNA synthetase containing no other acyl-*t*RNA synthetase was bound to CNBr-activated Sepharose, and used as an adsorbent. When histidyl-*t*RNA labeled with ^{3}H was passed through a column packed with the adsorbent, it was strongly adsorbed, and could be eluted by 1 N saline, as shown in Fig. 4.53.[240] Similarly, by using immobilized isoleucyl-*t*RNA synthetase as an adsorbent, isoleucyl-*t*RNA was isolated.[272]

As described above, affinity chromatography can be widely used for the purification of biochemicals such as enzymes, nucleic acids, enzyme inhibitors, substrates, reaction products, antibodies, hormones, and so forth.

The stability and capacity of the adsorbent and its cost are still problems, especially for large-scale affinity chromatography for industrial purposes. If these problems are overcome, in view of its specificity and

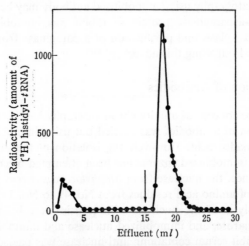

Fig. 4.53 Purification of histidyl-*t*RNA using immobilized histidyl-*t*RNA synthetase.[240]
The column was equilibrated with 66 mM potassium cacodylate, pH 7.0, containing 8 mM reduced glutathione and 10 mM MgCl$_2$. [^{3}H]-Histidyl-*t*RNA dissolved in the above buffer was applied to the column. Elution was carried out with 3.0 M NaCl containing acetic acid, pH 3.0; it was also possible to elute with 1 N NaCl (arrow).

efficiency, affinity chromatography may become very important industrially.

4.7 TREATMENT OF WASTE WATER

The application of immobilized enzymes and microbial cells for waste treatment has been studied.

For example, cyanide in waste water is usually decomposed by the action of microorganisms, by the so-called active sludge method. To improve this method, the cyanide-decomposing enzyme was immobilized by the entrapping method using polyacrylamide gel, packed into a column, and removal of cyanide from waste water using the resulting enzyme column was investigated.[273] Humphrey and his co-workers of the University of Pensylvania, U.S.A., studied the continuous measurement of specific substances in waste water by using immobilized enzymes, and also studied the reduction of phenol content in waste water by using immobilized polyphenol oxidase.[274]

Further, conversion of ε-aminocaproic acid cyclic dimer in waste water from a nylon plant to a bio-degradable substance was studied using *Achromobacter guttatus* immobilized in polyacrylamide gel.[275]

Though many kinds of harmful substances in waste water cannot be simultaneously decomposed, as they can in the case of the active sludge method, these systems are advantageous for the treatment and removal of specific substances. Thus, further development of systems utilizing immobilized enzymes and microbial cells in this field may be expected.

4.8 FUEL CELLS AND BIOCHEMICAL BATTERIES

Application of immobilized enzymes for fuel cells and biochemical batteries is potentially very interesting. The main reaction by which organisms obtain energy is oxidation of nutrients, and chemical energy produced by oxidative reactions is utilized to maintain life. Essentially, these reactions involve the conversion of electron-rich substances (nutrients) to electron-poor substances (metabolites). If a part of the electron-transfer system can be used for an electrode reaction, and chemical energy can be transformed to electrical energy, the manufacture of fuel cells or biochemical batteries may be possible.

A fuel cell in which uncharged *N*-acetyl–L-glutamate diamide is converted to ammonia and *N*-acetyl–L-glutamine by immobilized papain has been designed (Eq. 4.55).[276]

A fuel cell utilizing immobilized glucose oxidase has also been investigated.[277] This enzyme was immobilized by entrapping in polyacrylamide

$$H_2NOCCH_2CH_2CHCONH_2 + H_2O \xrightarrow{\text{papain}} H_2NOCCH_2CH_2CHCOOH + NH_4^+ + e^-$$
$$\qquad\qquad | \qquad\qquad\qquad\qquad\qquad\qquad\qquad\qquad | $$
$$\qquad NHCOCH_3 \qquad\qquad\qquad\qquad\qquad\qquad NHCOCH_3$$

(4.55)

gel, and an enzyme electrode was prepared by combining the resulting immobilized enzyme with a platinum electrode. Optimum conditions were investigated to obtain the highest voltage with gluconic acid and electrons produced from glucose according to Eq. 4.56.

$$glucose + O_2 \xrightarrow{\text{glucose oxidase}} gluconic\ acid + H^+ + e^-$$

(4.56)

In addition, immobilized enzyme was used for the conversion of solar energy to chemical energy.[278] That is, hydrogenase of *Clostridium pasteurianum* was immobilized by using carbodiimide and succinylaminoalkylated glass beads. This immobilized preparation could produce hydrogen using ferredoxin as a substrate on photo-irradiation in a chloroplast-ferredoxinhydrogenase system. Also, *Rhodospirillium rubrum* immobilized in a plate reactor could produce hydrogen using malate as a substrate; the amount of hydrogen produced from unit glucose and its cost were discussed.[279] Similarly, *Clostridium butyricum* immobilized in polyacrylamide gel produced hydrogen using glucose as a substrate, and application of the resulting immobilized preparation was studied as a biochemical battery.[280]

Immobilized enzymes are relatively stable during the conversion of chemical energy to electrical energy, and are easy to handle, so they are attractive as a possible new energy conversion system.

4.9 MISCELLANEOUS APPLICATIONS

Various other applications have also been reported for immobilized systems.

The cold sterilization of *Staphylococcus aureus, Escherichia coli*, and so on, was carried out by using peroxidase immobilized on CNBr-activated Sepharose in the presence of H_2O_2 and KI.[281]

Suzuki *et al.* investigated the application of immobilized enzymes for anti-pollution procedures. That is, air sterilization was carried out by using RNase immobilized on a collagen membrane.[282] Cold sterilization of water was attempted by using lytic enzyme immobilized on a collagen membrane.[283] These sterilization methods are promising especially for hospitals or pharmaceutical plants where air quality requirements are strict. Further, they carried out studies on the control of the activity of enzyme collagen membranes by external energy such as light.[284,285] Photo-sensitive enzyme was prepared by chemical modification using spiropyrane, a well-known photochromic compound. This enzyme preparation was immobilized on collagen membrane, and the activity of the immobilized preparation was monitored under ultraviolet and visible light and in the dark. The activity under ultraviolet light was 62% of that under visible light or in the dark. The activity of native enzyme was not affected by ultraviolet irradiation, and the effect may be explained in terms of conformational change of the enzyme due to ring closure of spiropyrane, as shown in Fig. 4.54.

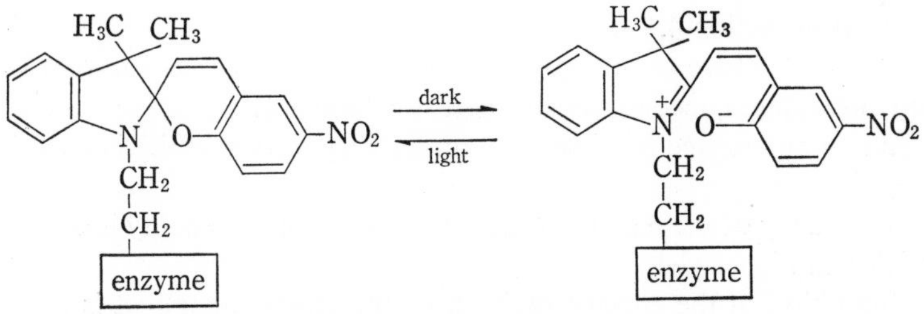

Fig. 4.54.   Spiropyrane-enzyme complex.

4.10   CONCLUSION AND PROSPECTS

In previous sections, we have briefly described applications of immobilized enzymes and immobilized microbial cells. Among these applications, most use is immobilized system essentially as solid catalysts for synthetic chemical reactions, and many such papers have appeared. However, examples of immobilized systems used as solid catalysts for industrial purposes are at present limited to the following 5 cases.

1)   Production of L-amino acids by the optical resolution of acetyl–DL-amino acids using immobilized aminoacylase (in Japan).

2)   Production of 6-amino penicillanic acid using immobilized penicillin amidase (in the U.S.A., Europe, and Japan).

3)   Production of fructose syrup using immobilized glucose isomerase (in the U.S.A., Europe, and Japan).

4)   Production of L-aspartic acid using immobilized microbial cells (in Japan).

5)   Production of L-malic acid using immobilized microbial cells (in Japan).

The limited number of industrial applications, in contrast with the large number of published reports, is due to the following problems.

1)   The carrier or reagent for immobilization is expensive.

2)   The yield of activity of immobilized enzyme per enzyme used is low, that is, the efficiency of immobilization is low.

3) The operational stability of immobilized enzymes is not yet good enough.

4) For continuous reaction, complicated equipment is required.

5) Equipment for continuous reaction is not multi-purpose.

6) Demand for the products is not sufficient to obtain significant advantages of scale.

7) Immobilized enzymes are less effective on high-molecular-weight substrates.

The reasons for loss of enzyme activity during operation have been described in section 4.1.4.

Although there are the above problems, immobilized enzymes have many advantages, as described in part in Chapter 1.

1) When an enzyme is adequately immobilized, its stability is enhanced in many cases.

2) The purity of the reaction product is high, and the quality is constant, resulting in higher yield.

3) Inhibition of the enzyme reaction by the reaction product is reduced in some cases.

4) The enzyme reaction is easily controlled, and reaction conditions can be held constant.

5) Automatic control is relatively easy, and labor requirements are reduced.

6) Shift of the optimum pH for reaction or a change of the kinetic constant is possible by appropriate immobilization.

7) Side reactions are diminished compared with the direct use of microbial cells for the reaction.

8) Immobilized enzyme is easily recovered after the enzyme reaction (by filtration, specific gravity separation, centrifugation, or magnetic separation), and can be reused.

9) Productivity per unit enzyme is enhanced.

10) Mass production is possible with compact equipment, i.e., a continuous process requires only a small amount of space for production.

Considering these advantages of immobilized enzymes, efficient combination with an energy-generating system or redox system would even further expand the fields of application. As described in the discussion on multi-enzyme systems, many such studies have already been carried out. If these systems are suitably developed, useful substances now produced by fermentation may be produced more efficiently and cheaply by combinations of immobilized enzymes.

As regards the use of immobilized microbial cells, industrial applications by the authors for the production of L-aspartic acid and L-malic acid have already been mentioned (*see* section 4.2.4). In these methods, isolation and extraction of the enzyme from microbial cells are unnecessary, and loss of the enzyme is minimized. This method cannot be utilized when a substrate or product is of high molecular weight, but can be used when enzymes catalyzing side reactions or decomposition reactions can be eliminated. Thus, direct immobilization of microbial cells for use as solid catalysts is likely to become more widespread, especially for the production of materials now produced by fermentation.

Academically, immobilized enzymes may also be used for various purposes, such as elucidation of complicated enzyme reaction mechanisms, as models of membrane-bound enzymes, for the separation and purification of biologically active substances, and so on.

In addition, further applications of immobilized enzymes in the medical area, especially for the diagnosis and therapy of various diseases, and in artificial organs are expected to develop. Specific and rapid analysis methods using immobilized enzymes are useful as an aid to exact diagnosis, and the development of enzyme microcapsules without accompanying side effects should contribute to the therapy of various diseases.

Industries utilizing biochemical reactions represent a major economic growth point. In the future, immobilized enzymes and immobilized microbial cells may be used extensively in the production of foods, antibiotics, vitamins, hormones, amino acids, and nucleic acids, as well as in medical applications and the treatment of industrial effluents.

REFERENCES

1. M. D. Lilly and P. Dunnill, *Process Biochem.*, **6** (8), 29 (1971)
2. T. Tosa, T. Mori, N. Fuse and I. Chibata, *Agr. Biol. Chem.* (Tokyo), **33**, 1047 (1969)
3. F. X. Hasselberger, B. Allen, E. K. Paruchuri, M. Charles and R. W. Coughlim, *Biochem. Biophys. Res. Commun.*, **57**, 1054 (1974)
4. G. Gellf and J. Boundrant, *Biochim. Biophys. Acta*, **334**, 467 (1974)
5. W. R. Vieth and K. Venkatasubramanian, *Chemtech*, **3**, 677 (1973)
6. J. Kozeny, *Sitzber. Akad. Wiss. Wien, Abt. IIa*, **136**, 271 (1927)
7. P. C. Carman, *Trans. Inst. Chem. Engrs.* (London), **15**, 150 (1937)
8. T. Tosa, T. Mori and I. Chibata, *J. Ferment. Technol.*, **49**, 522 (1971)
9. I. Chibata, T. Tosa and T. Sato, *Methods in Enzymology*, vol. 44, p. 739, Academic Press, 1976
10. T. Tosa, T. Sato, T. Mori, Y. Matuo and I. Chibata, *Biotechnol. Bioeng.*, **15**, 69 (1973)
11. T. Sato, T. Mori, T. Tosa, I. Chibata, M. Furui, K. Yamashita and A. Sumi *ibid.*, **17**, 1797 (1975)
12. K. Yamamoto, T. Tosa, K. Yamashita and I. Chibata, *Eur. J. Appl. Microbiol.*, **3**, 169 (1976)
13. K. Yamamoto, T. Tosa, K. Yamashita and I. Chibata, *Biotechnol. Bioeng.*, **19**, 1101 (1977)

14. H. H. Weetall and G. Baum, Abstr. Papers, Am., Chem. Soc., No. 158, Biol., 153 (1969)
15. J. R. Wykes, P. Dunnill and M. D. Lilly, *Nature*, 230, 187 (1971)
16. S. Fukui, S. Ikeda, M. Fujimura, H. Yamada and H. Kumagai, *Eur. J. Biochem.*, 51, 155 (1975)
17. S. Fukui, S. Ikeda, M. Fujimura, H. Yamada and H. Kumagai, *Eur. J. Appl. Microbiol.*, 1, 25 (1975)
18. K. Mosbach and P.-O. Larsson, *Biotechnol. Bioeng.*, 12, 19 (1970)
19. S. S. Sofer, D. M. Ziegler and R. P. Popovich, *Biochem. Biophys. Res. Commun.*, 57, 183 (1974)
20. S. Ogino, *Agr. Biol. Chem.* (Tokyo), 34, 1268 (1970)
21. D. L. Marshall and J. L. Walter, *Carbohydrate Res.*, 25, 489 (1972)
22. J. C. Smith, I. J. Stratford, D. W. Hutchinson and H. J. Brentnall, *FEBS Letters*, 30, 246 (1973)
23. C. H. Hoffman, E. Harris, S. Chodroff, S. Michelson, J. W. Rothrock, E. Peterson and W. Reuter, *Biochem. Biophys. Res. Commun.*, 41, 710 (1970)
24. H. G. Gassen and R. Nolte, *ibid.*, 44, 1410 (1971)
25. D. L. Marshall, *Biotechnol. Bioeng.*, 15, 447 (1973)
26. H. M. Walton, J. E. Eastman and A. E. Staley, *ibid.*, 15, 951 (1973)
27. T. A. Butterworth, D. I. C. Wang and A. J. Sinskey, *ibid.*, 12, 615 (1970)
28. J. J. Marshall and W. J. Whelan, *Chem. Ind.* (London), No. 25, 701 (1971)
29. I. Christison, *Chem. Ind.*, 4, 215 (1972)
30. R. J. H. Wilson and M. D. Lilly, *Biotechnol. Bioeng.*, 11, 349 (1969)
31. S. P. O'neill, P. Dunnill and M. D. Lilly, *ibid.*, 13, 337 (1971)
32. M. J. Bachler, G. W. Strandberg and K. L. Smiley, *ibid.*, 12, 85 (1970)
33. K. L. Smiley, *ibid.*, 13, 309 (1971)
34. S. Usami, M. Matsubara and J. Noda, *Hakkokyokaishi* (Japanese), 29, 195 (1971)
35. C. Gruesbeck and H. F. Rase, *Ind. Eng. Chem. Prod. Res. Develop.*, 11, 74 (1972)
36. H. Maeda, S. Miyamichi and H. Suzuki, *Hakkokyokaishi* (Japanese), 28, 391 (1970)
37. T. K. Ghose and J. A. Kostick, *Biotechnol. Bioeng.*, 12, 921 (1970)
38. M. Mandels, J. Kostick and R. Parizek, *J. Polymer Sci.*, part C, No. 36, 445 (1971)
39. R. D. Mason and H. H. Weetall, *Biotechnol. Bioeng.*, 14, 637 (1972)
40. H. Suzuki, Y. Ozawa and H. Maeda, *Agr. Biol. Chem.* (Tokyo), 30, 807 (1966)
41. S. Usami and Y. Kuratsu, *Hakkokogakuzasshi* (Japanese), 51, 789 (1973)
42. D. Dinelli, *Process Biochem.*, 7, 9 (1972)
43. H. Negoro, *J. Ferment. Technol.*, 48, 689 (1970)
44. R. Koelsch, *Enzymologia*, 42, 257 (1972)
45. N. Grubhofer and L. Schleith, *Naturwissenshaften*, 40, 508 (1953)
46. H. H. Weetall and R. D. Mason, *Biotechnol. Bioeng.*, 15, 455 (1973)
47. D. A. Self, G. Kay, M. D. Lilly and P. Dunnill, *ibid.*, 11, 337 (1969)
48. D. Y. Ryu, C. F. Bruno, B. K. Lee and K. Venkatasubramanian, Proceedings of the IVth International Fermentation Symposium: Fermentation Technology Today, p. 307, Society of Fermentation Technology, Japan, 1972
49. T. Kamogashira, T. Kawaguchi, W. Miyazaki and T. Doi, Japanese Patent, 72-28190 (1972)
50. Y. Ohno and M. Stahmann, *Macromolecules*, 4, 350 (1971)
51. T. Tosa, T. Mori, N. Fuse and I. Chibata, *Enzymologia*, 31, 214 (1966)
52. T. Tosa, T. Mori, N. Fuse and I. Chibata, *ibid.*, 31, 225 (1966)
53. T. Tosa, T. Mori, N. Fuse and I. Chibata, *ibid.*, 32, 153 (1967)
54. T. Barth and H. Mašková, *Collection Czech. Chem. Commun.*, 36, 2398 (1971)
55. T. Tosa, T. Mori, N. Fuse and I. Chibata, *Biotechnol. Bioeng.*, 9, 603 (1967)
56. T. Tosa, T. Mori and I. Chibata, *Agr. Biol. Chem.* (Tokyo), 33, 1053 (1969)
57. G. S. Omenn, D. A. Ontjes and C. B. Anfinsen, *Biochemistry*, 9, 313 (1970)
58. H. H. Weetall and C. C. Detar, *Biotechnol. Bioeng.*, 16, 1537 (1974)

59. M. A. Mitz and L. J. Summaria, *Nature*, **189**, 576 (1961)
60. M. D. Lilly, C. Money, W. E. Hornby and E. M. Crook, *Biochem. J.*, **95**, p. 45 (1965)
61. S. Kudo and H. Kushiro, Japanese Patent, 68-15401 (1968)
62. J. Kirimura and J. Yoshida, Japanese Patent, 64-27492 (1964)
63. M. Tanaka, N. Nakamura and K. Mineura, Japanese Patent, 68-26286 (1968)
64. Kyowa Hakko Kogyo Kabushiki Kaisha, France Patent, 1471792 (1966)
65. H. Mašková, T.Barth, B.Jirovský and I. Rychlik, *Collect. Czech. Chem. Commun.*, **38**, 943 (1973)
66. F. Leuschner, Ger. Patent, 1227855 (1966)
67. T. Fujishima and H. Yoshino, *Hakko to Taisha* (Japanese), **16**, 45 (1967)
68. W. Becker and E. Pfeil, *J. Am. Chem. Soc.*, **88**, 4299 (1966)
69. G. W. Strandberg and K. L. Smiley, *Biotechnol. Bioeng.*, **14**, 509 (1972)
70. N. Tsumura and M. Ishikawa, *Shokuhin Kogyo-Gakkaishi* (Japanese), **14**, 539 (1967)
71. G. W. Strandberg and K. L. Smiley, *Appl. Microbiol.*, **21**, 588 (1971)
72. J. C. Davis, *Chem. Eng.*, *Aug.*, **19**, p. 52, (1970)
73. N. H. Mermelstein, *Food Technol.*, **29**, 20 (1975)
74. H. H. Weetall and G. Baum, *Biotechnol. Bioeng.*, **12**, 399 (1970)
75. I. Chibata, T. Ishikawa and S. Yamada, *Bull. Agr. Chem. Soc. Japan*, **21**, 291 (1957)
76. I. Chibata, T. Ishikawa and S. Yamada, *ibid.*, **21**, 300 (1957)
77. I. Chibata, T. Ishikawa and S. Yamada, *ibid.*, **21**, 304 (1957)
78. I. Chibata, T. Tosa, T. Sato and T. Mori, *Methods in Enzymology*, vol. 44, p. 746, Academic Press, 1976
79. I. Chibata, T. Tosa, T. Sato, T. Mori and Y. Matuo, Proceedings of the IVth International Fermentation Symposium: Fermentation Technology Today, p. 383, Society of Fermentation Technology, Japan, 1972
80. T. Sato, T. Mori, T. Tosa and I. Chibata, *Arch. Biochem. Biophys.*, **147**, 788 (1971)
81. T. Mori, T. Sato, T. Tosa and I. Chibata, *Enzymologia*, **43**, 213 (1972)
82. T. Tosa, T. Mori and I. Chibata, *ibid.*, **40**, 49 (1971)
83. I. Chibata and T. Tosa, *Applied Biochemistry and Bioengineering*, vol. 1, p. 329, Academic Press, 1976
84. T. Kakimoto, T. Shibatani, N. Nishimura and I. Chibata, *Appl. Microbiol.*, **22**, 992 (1971)
85. K. Yamamoto, T. Sato, T. Tosa and I. Chibata, *Biotechnol. Bioeng.*, **16**, 1589 (1974)
86. I. Stone, U.S. Patent, 2717852 (1955)
87. H. H. Weetall, N. B. Havewala, W. H. Pitcher, Jr., C. C. Detar, W. P. Vann and S. Yaverbaum, *Biotechnol. Bioeng.*, **16**, 295 (1974)
88. H. H. Weetall, W. P. Vann, W. H. Pitcher, Jr., D. D. Lee, Y. Y. Lee and G. T. Tsao, *Methods in Enzymology*, vol. 44, p. 776, Academic Press, 1976
89. A. Kimura, H. Shirasaki and S. Usami, *Kogyokagakuzasshi* (Japanese), **72**, 489 (1969)
90. H. H. Weetall, *Process Biochem.*, **10**, 3 (1975)
91. S. P. O'Neill, J. R. Wykes, P. Dunnill and M. D. Lilly, *Biotechnol. Bioeng.*, **13**, 319 (1971)
92. J. Boudrant and C. Cheftel, *Biochimie*, **55**, 413 (1973)
93. S. Shaltiel, R. Mizrahi, Y. Stupp and M. Sela, *Eur. J. Biochem.*, **14**, 509 (1970)
94. T. Sato, T. Tosa and I. Chibata, *Eur. J. Appl. Microbiol.*, **2**, 153 (1976)
95. I. Chibata, T. Tosa and T. Sato, *Appl. Microbiol.*, **27**, 878 (1974)
96. T. Tosa, T. Sato, T. Mori and I. Chibata, *ibid.*, **27**, 886 (1974)
97. I. Chibata, T. Tosa and K. Yamamoto, *Enzyme Engineering*, 3, in press
98. T. Shibatani, N. Nishimura, K. Nabe, T. Kakimoto and I. Chibata, *Appl. Micro-*

biol., **27**, 688 (1974)
99. K. Yamamoto, T. Sato, T. Tosa and I. Chibata, *Biotechnol. Bioeng.*, **16**, 1601 (1974)
100. R. A. Messing, *ibid.*, **16**, 897 (1974)
101. Y. Fujita, Japanese Patent Kokai, 76-70871 (1976)
102. L. Zittan, P. B. Poulsen and S. H. Hemmingsen, *Die Stärk*, **27**, 236 (1975)
103. T. Kasumi, K. Kawashima and N. Tsumura, *Hakko Kogaku Zasshi* (Japanese), **51**, 321 (1974)
104. S. Glovenco, F. Morisi and P. Pansolli, *FEBS Letters*, **36**, 57 (1973)
105. N. B. Havewala and W. H. Pitcher, Jr., *Enzyme Engineering*, **2**, 315 (1974)
106. Y. Takasaki and A. Kanbayashi, *Kogyo Gijutsu Biseibutsu Kogyo Gijutsu Kenkyusho Hokoku* (Japanese), No. 37, 31 (1969)
107. M. B. Cohen, L. Spolter, C. C. Chang, N. S. MacDonald, J. Takahashi and D. D. Bobinet, *J. Nucl. Med.*, **15**, 1192 (1974)
108. A. Traub, E. Kaufmann and Y. Teitz, *Anal. Biochem.*, **28**, 469 (1969)
109. T. Fukumura, Japanese Patent, 74-15795 (1974)
110. H. D. Brown, A. B. Patel and S. K. Chattopadhyay, *J. Chromatog.*, **35**, 103 (1968)
111. P. A. Srere and K. Mosbach, *Annu. Rev. Microbiol.*, **28**, 61 (1974)
112. K. Mosbach and B. Mattiasson, *Acta Chem. Scand.*, **24**, 2084 (1970)
113. K. Mosbach and B. Mattiasson, *ibid.*, **24**, 2093 (1970)
114. R. Goldman and E. Katchalski, *J. Theoret. Biol.*, **32**, 243 (1971)
115. R. L. Lawrence and V. Okay, *Biotechnol. Bioeng.*, **15**, 217 (1973)
116. R. D. Falb, J. Lynn and J. Shapira, *Experientia*, **28**, 958 (1973)
117. M. D. Kamen and N. O. Kaplan, *Chem. Eng. News, Feb.*, 25, p. 19 (1974)
118. K. J. Skinner, *ibid.*, Aug. 18, p. 22 (1975)
119. W. Slowinski and S. E. Charm, *Biotechnol. Bioeng.*, **15**, 973 (1973)
120. S. Shimizu, H. Morioka, Y. Tani and K. Ogata, *J. Ferment. Technol.*, **53**, 77 (1975)
121. S. Yagi, Y. Toda and T. Minoda, Annual Meeting of the Agricultural Chemical Society of Japan, at Kyoto, April 4 (1976) p. 414
122. G. Corrieu, H. Blachere, A. Ramirez, J. M. Navarro, G. Durand, I. N. S. A. Toulouse, B. Duteurtre and M. Moll, Fifth International Fermentation Symposium, Berlin, 1976, p. 294
123. N. Takamatsu, K. Miyamoto, M. Okazaki and T. Miura, Annual Meeting of the Society of Fermentation Technology, Japan, at Osaka, October 30 (1973) p. 132
124. H. H. Weetall, *Anal. Chem.*, **46**, 602A (1974)
125. G. G. Guilbault, *ibid.*, **38**, 527R (1966)
126. G. G. Guilbault, *ibid.*, **40**, 459R (1968)
127. G. G. Guilbault, *Record Chem. Progr.*, **30**, 261 (1969)
128. G. G. Guilbault, *CRC Crit. Rev. Anal. Chem.*, **1**, 377 (1970)
129. D. A. Gough and J. D. Andrade, *Science*, **180**, 380 (1973)
130. W. E. Hornby, D. J. Inman and A. McDonald, *FEBS Letters*, **23**, 114 (1972)
131. G. P. Hicks and S. J. Updike, *Anal. Chem.*, **38**, 726 (1966)
132. D. J. Inman and W. E. Hornby, *Biochem. J.*, **129**, 255 (1972)
133. J. Campbell, W. E. Hornby and D. L. Morris, *Biochim. Biophys. Acta*, **384**, 307 (1975)
134. T. Tosa, R. Sano and I. Chibata, *Agr. Biol. Chem.* (Tokyo), **38**, 1529 (1974)
135. E. Riesel and E. Katchalski, *J. Biol. Chem.*, **239**, 1521 (1964)
136. D. J. Inman and W. E. Hornby, *Biochem. J.*, **137**, 25 (1974)
137. S. Ikeda, Y. Sumi and S. Fukui, *FEBS Letters*, **47**, 295 (1974)
138. S. Ikeda and S. Fukui, *ibid.*, **41**, 216 (1974)
139. D. R. Senn, P. W. Carr and L. N. Klatt, *Anal. Chem.*, **48**, 954 (1976)
140. A. Johansson, J. Lundberg, B. Mattiasson and K. Mosbach, *Biochim. Biophys. Acta*, **304**, 217 (1973)
141. K. Mosbach and B. Danielsson, *ibid.*, **364**, 140 (1974)

142. H.-L. Schmidt, G. Krisam and G. Grenner, *ibid.*, **429**, 283 (1976)
143. G. G. Guilbault and D. N. Kramer, *Anal. Chem.*, **37**, 1675 (1965)
144. S. J. Updike and G. P. Hicks, *Nature*, **214**, 986 (1967)
145. M. K. Weibel, W. Dritschilo, H. J. Bright and A. E. Humphrey, *Anal. Biochem.*, **52**, 402 (1973)
146. G. G. Guilbault and E. Hrabankova, *Anal. Chem.*, **42**, 1779 (1970)
147. G. G. Guilbault and G. Nagy, *Anal. Letters*, **6**, 301 (1973)
148. G. G. Guilbault and F. R. Shu, *Anal. Chim. Acta*, **56**, 333 (1971)
149. G. G. Guilbault and E. Hrabankova, *ibid.*, **56**, 285 (1971)
150. L. B. Wingard, Jr., C. C. Liu and N. L. Nagda, *Biotechnol. Bioeng.*, **13**, 629 (1971)
151. G. G. Guilbault and J. G. Montalvo, *J. Am. Chem. Soc.*, **92**, 2533 (1970)
152. G. J. Papariello, A. K. Mukherji and C. M. Shearer, *Anal. Chem.*, **45**, 790 (1973)
153. R. A. Llenado and G. A. Rechnitz, *ibid.*, **43**, 1457 (1971)
154. D. R. Senn, P. W. Carr and L. N. Klatt, *ibid.*, **48**, 954 (1976)
155. M. Aizawa, I. Karube and S. Suzuki, *Anal. Chim. Acta*, **69**, 431 (1974)
156. I. Satoh, I. Karube and S. Suzuki, *Biotechnol. Bioeng.*, **19**, 1095 (1977)
157. I. Satoh, I. Karube and S. Suzuki, *ibid.*, **18**, 269 (1976)
158. G. G. Guilbault and F. R. Shu, *Anal. Chem.*, **44**, 2161 (1972)
159. B. K. Ahn, S. K. Wolfson, Jr. and S. J. Yao, *Bioelectrochem. Bioenergetics*, **2**, 142 (1975)
160. P. Davies and K. Mosbach, *Biochim. Biophys. Acta*, **370**, 329 (1974)
161. J. F. Rusling, G. H. Luttrell, L. F. Cullen and G. J. Papariello, *Anal. Chem.*, **48**, 1211 (1976)
162. C. N. Statham, M. J. Melancon, Jr. and J. J. Lech, *Chem. Eng, News*, Oct. 25, p. 23 (1976)
163. I. Karube, S. Mitsuda, T. Matsunaga and S. Suzuki, *J. Ferment. Technol.*, **55**, 243 (1977)
164. H. H. Weetall and N. Weliky, *Anal. Biochem.*, **14**, 160 (1966)
165. R. O. Stasiw, A. B. Patel and H. D. Brown, *Biotechnol. Bioeng.*, **14**, 629 (1972)
166. E. Katchalski and A. Bar-Eli, *Nature*, **188**, 856 (1960)
167. J. J. Cebra, *J. Immunology*, **92**, 977 (1964)
168. G. P. Royer and G. M. Green, *Biochem. Biophys. Res. Commun.*, **44**, 426 (1971)
169. G. P. Royer and J. P. Andrews, *J. Biol. Chem.*, **248**, 1807 (1973)
170. H. P. J. Bennett, D. F. Elliot, B. E. Evans, P. J. Lowry and C. McMartin, *Biochem. J.*, **129**, 695 (1972)
171. D. Gabel and B. V. Hofsten, *Eur. J. Biochem.*, **15**, 410 (1970)
172. C. Schwabe, *Biochemistry*, **8**, 795 (1969)
173. S. Lowey, L. Goldstein, C. Cohen and S. M. Luck, *J. Mol. Biol.*, **23**, 287 (1967)
174. R. A. Zingaro and M. Uziel, *Biochim. Biophys. Acta*, **213**, 371 (1970)
175. J. C. Lee, *ibid.*, **235**, 435 (1971)
176. D. G. Knorre, N. V. Melamed, V. K. Starostina and T. N. Shubina, *Biochemistry (USSR) (English Transl.)*, **38**, 101 (1973)
177. A. Rimon, B. Alexander and E. Katchalski, *Biochemistry*, **5**, 792 (1966)
178. R. M. Howell and R. J. Dupe, *Biochem. J.*, **123**, 11p (1971)
179. W. G. Owen and R. H. Wagner, *Am. J. Physiol.*, **220**, 1941 (1971)
180. K. S. Stenn and E. R. Blout, *Biochemistry*, **11**, 4502 (1972)
181. A. Rimon, M. Gutman and S. Rimon, *Biochim. Biophys. Acta*, **73**, 301 (1963)
182. M. Gutman and A. Rimon, *Can. J. Biochem.*, **42**, 1339 (1964)
183. C. M. Ambrus, J. L. Ambrus, O. A. Roholt, B. K. Meyer and R. R. Shields, *J. Med.*, **3**, 270 (1972)
184. D. G. Deutsch and E. T. Mertz, *ibid.*, **3**, 224 (1972)
185. W. Brümmer, N. Hennrich, M. Klockow, H. Lang and H. D. Orth, *Eur. J. Biochem.*, **25**, 129 (1972)
186. E. B. Ong, Y. Tsang and G. E. Perlmann, *J. Biol. Chem.*, **241**, 5661 (1966)

187. T. Seki, T. A. Jenssen, Y. Levin and E. G. Erdös, *Nature*, **225**, 864 (1970)
188. S. Gestrelius, B. Mattiasson and K. Mosbach, *Eur. J. Biochem.*, **36**, 89 (1973)
189. P. A. Srere, B. Mattiasson and K. Mosbach, *Proc. Nat. Acad. Sci.*, **70**, 2534 (1973)
190. R. Goldman, H. I. Silman, S. R. Caplan, O. Kedem and E. Katchalski, *Science*, **150**, 758 (1965)
191. R. Goldman, O. Kedem, I. H. Silman, S. R. Caplan and E. Katchalski, *Biochemistry*, **7**, 486 (1968)
192. R. Goldman, O. Kedem and E. Katchalski, *ibid.*, **7**, 4518 (1968)
193. J. H. Julliard, C. Godinot and D. C. Gautheron, *FEBS Letters*, **14**, 185 (1971)
194. K. Feldmann, H. Zeisel and E. Helmreich, *Proc. Nat. Acad. Sci.*, **69**, 2278 (1972)
195. B. Mattiasson and K. Mosbach, *Biochim. Biophys. Acta*, **235**, 253 (1971)
196. W. W.-C. Chan, *Biochem. Biophys. Res. Commun.*, **41**, 1198 (1970)
197. W. W.-C. Chan and H. M. Mawer, *Arch. Biochem. Biophys.*, **149**, 136 (1972)
198. S. Ikeda and S. Fukui, *Eur. J. Biochem.*, **46**, 553 (1974)
199. D. A. Fell and C. J. B. White, *Biochem. Biophys. Res. Commun.*, **67**, 1013 (1975)
200. E. A. Wider de Xifra, S. Mendiara and A. M. delC Batlle, *FEBS Letters*, **27**, 275 (1972)
201. T. M. S. Chang, *Science*, **146**, 524 (1964)
202. T. M. S. Chang and M. J. Poznansky, *Nature*, **218**, 243 (1968)
203. M. J. Poznansky and T. M. S. Chang, *Proc. Can. Fed. Biol. Soc.* **12**, 54 (1969)
204. T. M. S. Chang, *ibid.*, **12**, 62 (1969)
205. T. M. S. Chang, *Nature*, **229**, 117 (1971)
206. T. M. S. Chang, *Enzyme*, **14**, 95 (1973)
207. T. Mori, T. Tosa and I. Chibata, *Biochim. Biophys. Acta*, **321**, 653 (1973)
208. S. J. Updike, R. T. Wakamiya and E. N. Lightfoot, Jr., *Science*, **193**, 681 (1976)
209. T. Mori, T. Tosa and I. Chibata, *Cancer Res.*, **34**, 3066 (1974)
210. J. P. Allison, L. Davidson, A. G. Hartman and G. B. Kitto, *Biochem. Biophys. Res. Commun.*, **47**, 66 (1972)
211. D. Sampson, L. S. Hersh, D. Cooney and G. P. Murphy, *Trans. Am. Soc. Antificial Internal Organs*, **18**, 54 (1972)
212. Y. Inada, S. Hirose, M. Okada and H. Mihama, *Enzyme*, **20**, 188 (1975)
213. J. C. Venter, B. R. Venter, J. E. Dixon and N. O. Kaplan, *Biochem. Med.*, **12**, 79 (1975)
214. R. E. Sparks, O. Lindan and M. H. Litt, Annual Progress Report, Chemical Engineering Science Division, Case Western Reserve Univ., Cleveland, Ohio, 1968
215. T. M. S. Chang, *Trans. Am. Soc. Antificial Internal Organs*, **12**, 13 (1966)
216. T. M. S. Chang, L. J. Johnson and O. J. Ransome, *Can. J. Physiol. Pharmacol.*, **45**, 705 (1967)
217. G. Broun, C. Tran-Mink, D. Thomas, D. Domurado and E. Selegny, *Trans. Am. Soc. Artificial Internal Organs*, **17**, 341 (1971)
218. D. L. Gardner, R. D. Falb, B. C. Kim and D. C. Emmerling, *ibid.*, **17**, 239 (1971)
219. G. Gregoriadis, P. D. Leathwood and B. E. Ryman, *FEBS Letters*, **14**, 95 (1971)
220. A. Dahlqvist, B. Mattiasson and K. Mosbach, *Biotechnol. Bioeng.*, **15**, 395 (1973)
226. L. E. Wierzbicki, V. H. Edwards and F. V. Kosikowsky, *J. Food Sci.*, **38**, 1070 (1973)
222. E. S. Okos and W. J. Harper, *ibid.*, **39**, 88 (1974)
223. L. E. Wierzbicki, V. H. Edwards and F. V. Kosikowsky, *ibid.*, **39**, 374 (1974)
224. K. Ohmiya, C. Terao, S. Shimizu and T. Kobayashi, *Agr. Biol. Chem.* (Tokyo) **39**, 491 (1975)
225. H. Samejima and K. Kimura, *Enzyme Engineering*, **2**, 131 (1974)
226. F. Morisi, M. Pastore and A. Viglia, *J. Dairy Sci.*, **56**, 1123 (1973)
227. H. Maeda, *Biotechnol. Bioeng.*, **17**, 1571 (1975)
228. M. L. Green and G. Crutchfield, *Biochem. J.*, **115**, 183 (1969)

229. H. R. Schreiner, U. S. Patent, 3282702 (1966)
230. W. F. Shipe, G. Senyk and H. H. Weetall, *J. Dairy Sci.*, **55**, 647 (1972)
231. P. R. Witt, Jr., R. A. Sair, T. Richardson and N. F. Olson, *Brewers Dig.*, October, p. 70 (1970)
232. L. Goldstein, A. Lifshitz and M. Sokolovsky, *Int. J. Biochem.*, **2**, 448 (1971)
233. F. Leuschner, British Patent, 953414 (1964)
234. H. H. Weetall and C. C. Detar, *Biotechnol. Bioeng.*, **16**, 1095 (1974)
235. N. W. H. Cheethan and G. N. Richards, *Carbohydrate Res.*, **30**, 99 (1973)
236. J. H. Reynolds, *Biotechnol. Bioeng.*, **16**, 135 (1974)
237. P. Cuatrecasas, *J. Agr. Food Chem.*, **19**, 600 (1971)
238. F. Friedberg, *Chromatog. Rew.*, **14**, 121 (1971)
239. P. Cuatrecasas and C. B. Anfinsen, *Ann. Rev. Biochem.*, **40**, 259 (1971)
240. P. Cuatrecasas and C. B. Anfinsen, *Methods in Enzymology*, vol. 22, p. 345, Academic Press, 1971
241. H. H. Weetall, *Separation Methods*, **2**, 199 (1973)
242. P. Cuatrecasas, *Advances in Enzymology*, vol. 36, p. 29, Wiley, 1972
243. M. Yamazaki, S. Ishii and K. Iwai, *Affinity Chromatography* (Japanese), Kodansha, 1975
244. I. Chibata, T. Tosa and Y. Matuo, *Affinity Chromatography—Zikken to Ōyo* (Japanese), Kodansha, 1976
245. R. Axén, J. Porath and S. Ernback, *Nature*, **214**, 1302 (1967)
246. E. Steers, P. Cuatrecasas and H. B. Polland, *J. Biol. Chem.*, **246**, 196 (1971)
247. K. Hochstrasser, M. Muss and E. Werle, *Hoppe-Seyler's Z. Physiol. Chem.*, **348**, 1337 (1967)
248. H. Fritz, M. Hutzel and E. Werle, *ibid.*, **348**, 950 (1967)
249. H. Fritz, H. Schult, M. Hutzel and M. Wiedemann, *ibid.*, **348**, 308 (1967)
250. K. K. Stewart and R. F. Doherty, *FEBS Letters*, **16**, 226 (1971)
251. M. U. Lindberg and L. Skoog, *Eur. J. Biochem.*, **13**, 326 (1970)
252. L. Sundberg and T. Kristiansen, *FEBS Letters*, **22**, 175 (1972)
253. E. Starkenstein, *Biochem. Z.*, **24**, 210 (1910)
254. S. Schwimmer and A. K. Balls, *J. Biol. Chem.*, **179**, 1063 (1949)
255. L. S. Lerman, *Proc. Nat. Acad. Sci.*, **39**, 232 (1953)
256. M. S. Poonian, A. J. Schlabach and A. Weissbach, *Biochemistry*, **10**, 424 (1971)
257. L. F. Cavalieri and E. Carroll, *Proc. Nat. Acad. Sci.*, **67**, 807 (1970)
258. G. K. Chua and W. Bushuk, *Biochem. Biophys. Res. Commun.*, **37**, 545 (1969)
259. T. Tosa, T. Sato, R. Sano, K. Yamamoto, Y. Matuo and I. Chibata, *Biochim. Biophys. Acta*, **334**, 1 (1974)
260. K. Aspberg and J. Porath, *Acta Chem. Scand.*, **24**, 1839 (1970)
261. J. R. Uren, *Biochim. Biophys. Acta*, **236**, 67 (1971)
262. C. R. Lowe and P. D. G. Dean, *FEBS Letters*, **14**, 313 (1971)
263. P. Cuatrecasas, M. Wilchek and C. B. Anfinsen, *Proc. Nat. Acad. Sci.*, **61**, 636 (1968)
264. G. Feinstein, *FEBS Letters*, **7**, 353 (1970)
265. M. Claeyssens, H. K.-Hilderson, J.-P. Vanwauwe and C. K. DeBruyne, *ibid.*, **11**, 336 (1970)
266. W. W. C. Chan and M. Takahashi, *Biochem. Biophys. Res. Commun.*, **37**, 272 (1969)
267. B. Sprossler and F. Lingens, *FEBS Letters*, **6**, 232 (1970)
268. I. Chibata, T. Tosa and Y. Matuo, *Enzyme Engineering*, **2**, 229 (1974)
269. R. Collier and G. Kohlhaw, *Anal. Biochem.*, **42**, 48 (1971)
270. H. H. Weetall, *Biochem. J.*, **117**, 257 (1970)
271. G. S. Omenn, D. A. Ontjes and C. B. Anfinsen, *Nature*, **225**, 189 (1970)
272. J. Denburg and M. Deluca, *Proc. Nat. Acad. Sci.*, **67**, 1057 (1970)
273. K. Kawai, K. Suga, T. Imanaka and H. Taguchi, Annual Meeting of the Agricultural Chemical Society of Japan, p. 159 (1971)

264 APPLICATIONS OF IMMOBILIZED SYSTEMS

274. A. E. Humphrey and E. K. Pye, *Chem. Eng. News*, Jan. 3, 25 (1972)
275. S. Kinoshita, M. Muranaka and H. Okada, *J. Ferment. Technol.*, 53, 223 (1975)
276. R. Blumenthal, S. R. Caplan and O. Kedem, *Biophys. J.*, 7, 735 (1967)
277. L. B. Wingard, Jr., C. C. Liu and N. L. Nagda, *Biotechnol. Bioeng.*, 13, 629 (1971)
278. D. A. Lappi, F. E. Stolzenback and N. O. Kaplan, *Biochem. Biophys. Res. Commun.*, 69, 878 (1976)
279. M. A. Bennett and H. H. Weetall, *J. Solid-Phase Biochem.*, 1, 137 (1976)
280. I. Karube, K. Matsunaga, S. Tsuru and S. Suzuki, *Biochim. Biophys. Acta*, 444, 338 (1976)
281. S. Henry, J. Koczan and T. Richardson, *Biotechnol. Bioeng.*, 16, 289 (1974)
282. I. Karube, T. Suganuma and S. Suzuki, *J. Ferment. Technol.*, 54, 751 (1976)
283. S. Suzuki and I. Karube, *Bōkin Bōbai Shi* (Japanese), 4, 111 (1976)
284. I. Karube, Y. Nakamoto, K. Namba and S. Suzuki, *Biochim. Biophys. Acta*, 429, 975 (1976)
285. Y. Nakamoto, I. Karube, S. Terawaki and S. Suzuki, *J. Solid-Phase Biochem.*, 1, 143 (1976)

Appendix Table

APPENDIX TABLE 1   Commercially available water-insoluble carriers

Kind of carrier	Substituent groups	Remarks	Manufacturer[†1]
A. AGAROSE SERIES			
Sepharose 2B	unsubstituted (OH)	40×10^{6}[†2] agarose 2%	Pharmacia
" 4B		20×10^{6}[†2] agarose 4%	
" 6B		4×10^{6}[†2] agarose 6%	
Bio-Gel A-150 m	"	150×10^{6}[†2]	Bio Rad
" A-50 m	"	50×10^{6}[†2]	
" A-15 m	"	15×10^{6}[†2]	
" A-5 m	"	5×10^{6}[†2]	
" A-1.5 m	"	1.5×10^{6}[†2]	
" A-0.5 m	"	0.5×10^{6}[†2]	
Chromagel A-2	"	40×10^{6}[†2] agarose 2%	Dozin
" A-4	"	20×10^{6}[†2] agarose 4%	
" A-6	"	4×10^{6}[†2] agarose 6%	
AH-Sepharose 4B	$-OCONH_2(CH_2)_6NH_2$		Pharmacia
Agarose-aminohexane	"		P-L Biochem.
Affi-Gel 101	$\sim NH_2$	spacer length 0.6 nm	Bio Rad
" 102	"	" 1 m	"
CH-Sepharose 4B	$-OCONH(CH_2)_6COOH$		Pharmacia
Agarose capronic acid	$-OCONH(CH)_6COOH$		P-L Biochem.
Affi Gel 201	$\sim COOH$	spacer length 1 nm	Bio Rad
" 202	"	" 1.5 nm	"
Agarose hexanethiol	$\sim SH$		P-L Biochem.

(Continued)

APPENDIX TABLE 1—*Continued* (1)

Kind of carrier	Substituent groups	Remarks	Manufacturer[†1]
Affi-Gel 401	∼SH	spacer length 1.5 nm	Bio Rad
" 501	organic mercurichloride	" 1.5 nm	Miles
PL-Agarose	poly-L-lysine-	multivalent spacer gel	"
PAL-Agarose	poly-DL-alanine-poly lysine-	"	"
SPL-Agarose	succinyl-poly-L-lysine-	"	"
SPAL-Agarose	succinyl-poly-DL-alanine-poly lysine-	"	"
CNBr-Activated Sepharose 4B	$-O-\!\!>\!C{=}NH$ ($-O-$)		Pharmacia
" 6MB	" OH	purification of cells	"
Epoxy-activated Sepharose 6B	$-OCH_2CHCH_2O(CH_2)_4OCH_2CH\text{-}CH_2$		"
Activated CH-Sepharose 4B	$-OCONH(CH_2)_6COON$		Pharmacia
Activated thiol Sepharose 4B	$-OCONHCH(CH_2)_2CONHCHCH_2S\text{-}S\text{-}N$ / $CONHCH_2COOH$		"
Agarose thiol coupler	$-S\text{-}S\text{-}N$		P-L Biochem.
Agarose-N-hydroxy-succinamide coupler	$-NH(CH_2)_3NH(CH_2)_3NHCO(CH_2)_2COO\text{-}N$		"
Affi-Gel 10	"		Bio Rad
Adipic acid hydrazide-agarose			Miles
Succinyladipic acid hydrazide-agarose	"		"

(*Continued*)

APPENDIX TABLE 1—*Continued* (2)

Kind of carrier	Substituent groups	Remarks	Manufacturer[†1]
DEAE–Sepharose CL 6B	$-OCH_2H_4\overset{+}{N}H(C_2H_5)_2$		Pharmacia
CM–Sepharose CL 6B	$-OCH_2COOH$		"
B. DEXTRAN SERIES			
Sephadex G-10	unsubstituted (OH)	bed volume[†3] 1.0	Pharmacia
" G-15	"	" 1.5	"
" G-25	"	" 2.5	"
" G-50	"	" 5.0	"
" G-75	"	" 7.5	"
" G-100	"	" 10	"
" G-150	"	" 15	"
" G-200	"	" 20	"
DEAE–Sephadex A-25	$-OC_2H_4\overset{+}{N}H(C_2H_5)_2$		"
" A-50	"		"
QAE–Sephadex A-25	$-OC_2H_4\overset{+}{N}(C_2H_5)_2CH_2CH(OH)CH_3$		"
" A-50	"		"
CM–Sephadex C-25	$-OCH_2COOH$		"
" C-50	"		"
SP–Sephadex C-25	$-O(CH_2)_3-SO_3H$		"
" C-50	"		"
C. CELLULOSE SERIES			
Cellex N-1	unsubstituted (OH)	fibers	Bio Rad
" MX	"	microcrystalline	"
" 410	"		"
CF-1	"	fibers	Whatman
CF-2	"		"
CC-31	"	microcrystalline	"
Avicel	"		Asahi Kasei

(Continued)

APPENDIX TABLE 1—*Continued* (3)

Kind of carrier	Substituent groups unsubstituted (OH)	Remarks	Manufacturer[1]
Cellulose powder (A)	unsubstituted (OH)	fibers	Toyo Roshi
(B)	"	"	"
(C)	"	"	"
Cellulose	"	microcrystalline	Merck
AE-cellulose	$-OCH_2CH_2NH_2$		Seravac
Cellex AE	"		Bio Rap
CM-cellulose	$-OCH_2COOH$		Seravac
Cellex-CM	"		Bio Rad
PAB-cellulose	$-OCH_2\!-\!\!\langle\!\!\!\bigcirc\!\!\!\rangle\!\!-NH_2$		Seravac
Cellex-PAB	"		Bio Rad
DEAE-cellulose	$-O(CH_2)_2\overset{+}{N}H_2(C_2H_{5/2})$		Seravac
ECTEOLA-cellulose			
P-cellulose	$-OPO_3H_2$		"
TEAE-cellulose	$-O(CH_2)_2NX(C_2H_5)_3$		"
SE-cellulose	$-O(CH_2)_2SO_3H$		"
PEI-cellulose	polyethyleneimine		"
Aminodecyl-cellulose	$-OCH_2CONH(CH_2)_{12}NH_2$		Merck
Aminohexyl-cellulose	$-OCH_2CONH(CH_2)_6NH_2$		"
Aminohexyl succinyl-cellulose	$-OH_2CONH(CH_2)_6\!-\!HNCO(CH_2)_2COOH$		"
CM-cellulose hydrazide	$-OCH_2CONHNH_2$		"

(Continued)

APPENDIX TABLE 1—*Continued* (4)

Kind of carrier	Substituent groups	Remarks	Manufacturer[1]
D. POLYACRYLAMIDE SERIES			
Bio-Gel P-2	unsubstituted ($-CONH_2$)	$2.6\times10^{3\dagger2}$	Bio Rad
" P-4	"	$3.6\times10^{3\dagger2}$	"
" P-6	"	$4.6\times10^{3\dagger2}$	"
" P-10	"	$12\times10^{3\dagger2}$	"
" P-30	"	$30\times10^{3\dagger2}$	"
" P-60	"	$60\times10^{3\dagger2}$	"
" P-100	"	$100\times10^{3\dagger2}$	"
" P-150	"	$150\times10^{3\dagger2}$	"
" P-200	"	$200\times10^{3\dagger2}$	"
" P-300	"	$300\times10^{3\dagger2}$	"
Chromagel P-1	"		Dozin
" P-2	"		"
" P-3	"		"
Bio-Gel P hydrazide	$-CONHNH_2$		Bio Rad
Enzafix P-HZ	$-CONHNH_2$		Wako
Enzacryl AH	"		Koch-Light
" AA	$-CONH-$⬡$-NH_2$		"
" PT	$-CONHCH(COOH)CH_2SH$		"
Enzacryl PTL	$-CONHCH\langle \begin{smallmatrix} CO-S \\ CH_2 \end{smallmatrix} \rangle$		Koch-Light
" PA	$-CONHCH_2CH(OCH_3)_2$		
AE-Bio-Gel P	$-CONH(CH_2)_2NH_2$		Bio Rad
Bio-Gel CM	$-COOH$		"
Enzafix P-SH	$-CONHNHCO(CH_2)_2COOH$		Wako
" P-AB	$-CONH(CH_2)_2NHCO-$⬡$-NH_2$		"

(*Continued*)

APPENDIX TABLE 1—*Continued* (5)

Kind of carrier	Substituent groups	Remarks	Manufacturer[t1]
E. GLASS SERIES			
Aminopropyl CPG			Elector-Nucleonics
Carboxyl CPG			"
CPG/dextran	$-(CH_2)_3NHCONH-O\rangle-OH$ $\rangle-OH$		Corning
CPG/long alkylamine	$-R(CH_2)_6NH_2$		"
CPG/phenylhydrazine	$-(CH_2)_3NHCO-\langle\rangle-NHNH_2\cdot HCl$		"
CPG/N-hydroxy succiinimidyl ester	$-(CH_2)_6NHCO(CH_2)_2COOR$		"
CPG/carboxyl	$-(CH_2)_3NHCO(CH_2)_2COOH$ $\underset{\;}{NHCOCH_3}$		"
CPG/thiol	$-(CH_2)_3NHCOCH_2CH_2SH$		"
CPG/p-nitrophenyl	$-(CH_2)_3NHCO-\langle\rangle-NO_2$		"
CPG/p-nitroaryl	$-(CH_2)_3NHCO-\langle\rangle-NO_2$		"
CPG/p-aminoaryl	$-(CH_2)_3NHCO-\langle\rangle-NH_2$		Corning

[t1] Abbreviations of manufacturers' names are summarized in Appendix Table 3.
[t2] Nominal molecular weight cut-off.
[t3] Bed volume (ml/g dry gel).

APPENDIX TABLE 2   Commercially available immobilized enzymes

Immobilized enzyme	Carrier	Manufacturer†
Alcohol dehydrogenase	agarose	Miles
//	DEAE-cellulose	Sigma
//	polyacrylamide	//
Glucose oxidase	agarose	Worthington
//	DEAE-cellulose	Miles
//	CM-cellulose	N.B.C.
//	polyacrylamide	Sigma
//	ambiguous	Boehringer
Glucose-6-phosphate dehydrogenase	agarose	Sigma
//	polyacrylamide	//
Malate dehydrogenase	//	//
Lactate dehydrogenase	//	//
Glyceraldehyde dehydrogenase	//	//
Cytochrome c	CM-cellulose	N.B.C.
Catalase	nitrocellulose	Worthington
Peroxidase	agarose	Sigma
//	//	Worthington
//	CM-cellulose	Miles
Hexokinase	polyacrylamide	Sigma
Pyruvate kinase	//	//
3-Phosphoglycerate kinase	//	//
Lipase	copolymer of ethylene and maleic acid	Monsanto
Cholinesterase	agarose	Sigma
Ficin	cellulose	//
//	CM-cellulose	Miles
//	//	Merck
Alkaline phosphatase	agarose	Sigma
//	//	Worthington
//	ambiguous	Boehringer
Ribonuclease	agarose	Miles
//	//	Sigma
//	//	Worthington
//	CM-cellulose	Merck
//	//	Miles
//	//	N.B.C.
//	polyacrylamide	Sigma
//	ambiguous	Boehringer
//	polymer of maleic acid	Merck
Deoxyribonuclease	agarose	Worthington
α-Amylase	cellulose	Miles
//	polyacrylamide	Sigma
//	copolymer of ethylene and maleic acid	Monsanto
β-Amylase	cellulose	Miles
Glucoamylase	polyacrylamide	Boehringer

(Continued)

APPENDIX TABLE 2—*Continued*

Immobilized enzyme	Carrier	Manufacturer†
Neuraminidase	agarose	Sigma
Glucuronidase	//	//
β-Galactosidase	nitrocellulose	Worthington
Leucine aminopeptidase	DEAE-cellulose	Miles
Carboxypeptidase	agarose	Sigma
//	//	Worthington
//	DEAE-cellulose	Miles
Prolidase	//	//
//	agarose	//
Chymotrypsin	//	Boehringer
//	//	Miles
//	//	Worthington
//	cellulose	Sigma
//	CM-cellulose	Merck
//	//	Miles
//	//	N.B.C.
//	copolymer of ethylene and maleic acid	Miles
Trypsin	agarose	//
//	//	Worthington
//	cellulose	//
//	CM-cellulose	Merck
//	//	Miles
//	//	N.B.C.
//	polyacrylamide	Sigma
//	copolymer of ethylene and maleic acid	Miles
//	polymer of maleic acid	Merck
//	ambiguous	Boehringer
Pronase	CM-cellulose	Merck
Protease	agarose	Miles
//	cellulose	Sigma
//	CM-cellulose	Miles
//	copolyer of ethylene and maleic acid	Monsanto
Subtilo peptidase A	copolymer of ethylene and maleic acid	Miles
Subtilo peptidase B	//	//
Subtilisin	CM-cellulose	Merck
Papain	agarose	Miles
//	cellulose	Sigma
//	CM-cellulose	Merck
//	//	Miles
//	//	N.B.C.
//	copolymer of ethylene and maleic acid	Miles
//	//	Monsanto

(*Continued*)

APPENDIX TABLE 2—*Continued*

Immobilized enzyme	Carrier	Manufacturer[†]
//	polymer of maleic acid	Merck
//	ambiguous	Boehringer
Proteinase	CM-cellulose	Merck
Bromelain	//	//
//	//	N.B.C.
//	ambiguous	Miles
Pepsin	AE-cellulose	Worthington
Urease	DEAE-cellulose	Miles
//	polyacrylamide	Sigma
//	ambiguous	Boehringer
Aminoacylase	ambiguous	Boehringer
Asparaginase	agarose	Sigma
Aldolase	polyacrylamide	//
Glucose phosphate isomerase	//	//
Glyceraldehyde-3-phosphate dehydrogenase and 3-phospho-glycerate phosphokinase	//	//
α-Glycerate dehydrogenase and triphosphate isomerase	//	//
Creatine phosphate kinase, hexokinase and glucose-6-phosphate dehydrogenase	//	//
Glyceraldehyde-3-phosphate dehydrogenase and 3-phospho-glycerate phosphokinase	//	//
Hexokinase and glucose-6-phosphate dehydrogenase	· agarose	//

† Abbreviations of manufacturers' name are summarized in Appendix Table 3.

APPENDIX TABLE 3 Abbreviations of manufacturers' names

Abbreviation	Manufacturer
Asahi Kasei	Asahi Kasei Kogyo Co. Ltd. (Japan)
Bio Rad	BIO-RAD Laboratories (U.S.A.)
Boehringer	Boehringer-Mannheim GmbH (Germany)
Corning	Corning Glass Works (U.S.A)
Dojin	Dojin Iyaku Kako Co. Ltd. (Japan)
Electro-Nucleonics	Electro-Nucleonics Laboratories Inc. (U.S.A.)
Koch-Light	Koch-Light Laboratories Ltd. (England)
Merck	E. Merck Darmstadt GmbH (Germany)
Miles	Miles-Seravac Laboratories Ltd. (England)
Monsanto	Monsanto Company (U.S.A.)
N.B.C.	Nutritional Biochemicals Corporation (U.S.A.)
Pharmacia	Pharmacia Fine Chemicals (Sweden)
Toyo Roshi	Toyo Roshi Co. Ltd. (Japan)
Wako	Wako Pure Chemical Industries Ltd. (Japan)

Abbreviations

A	Adenosine
Ac–Ala–Ala–Ala–ME	α-*N*-Acetyl–alanyl–alanyl–alanine methyl ester
Ac–DL-Met	α-*N*-Acetyl-DL-methionine
ADP	Adenosine 5′-diphosphate
AE-cellulose	Aminoethyl cellulose
L-Ala	L-Alanine
AMP	Adenosine 5′-phosphate
6-APA	6-Aminopenicillanic acid
L-Arg	L-Arginine
L-Asn	L-Asparagine
L-Asp	L-Aspartic acid
ATA	α-*N*-Acetyl-L-tyrosine amide
ATEE	α-*N*-Acetyl-L-tyrosine ethyl ester
ATP	Adenosine 5′-triphosphate
BAA	α-*N*-Benzoyl-L-arginine amide
BAEE	α-*N*-Benzoyl-L-arginine ethyl ester
BA*p*NA	α-*N*-Benzoyl-L-arginine-*p*-nitroanilide
BIS	*N*,*N*′-Methylene bisacrylamide
BT*p*NA	α-*N*-Benzoyl tyrosine-*p*-nitroanilide
BOD	Biochemical oxygen demand
Bz-Gly-L-Phe	α-*N*-Benzoyl glycine-L-phenylalanine
CDP	Cytidine 5′-diphosphate
L-Cit	L-Citrulline
CM-cellulose	Carboxymethyl cellulose
2′,3′-cCMP	Cytidine 2′,3′-cyclic phosphate
CoA	Coenzyme A
L-Cys	L-Cysteine
DAPA	Diaminopimelic acid
DEAA-cellulose	Diethylaminoacetyl cellulose
DEAE-cellulose	Diethylaminoethyl cellulose
DNA	Deoxyribonuleic acid
L-DOPA	L-Dihydroxyphenylalanine
ECTEOLA-cellulose	Epichlorohydrin triethanolamine cellulose
FDP	Fructose diphosphate
FMN	Flavin mononucleotide

G	Guanosine
GAP	Glyceroaldehyde-3-phosphate
Glc	Glucose
L-Glu	L-Glutamic acid
G-1-P	Glucose-1-phosphate
G-6-P	Glucose-6-phosphate
L-His	L-Histidine
IDP	Inosine 5'-diphosphate
IMP	Inosine 5'-phosphate
LEE	L-Lysine ethyl ester
L-Leu	L-Leucine
L*p*NA	L-Leucine-*p*-nitroanilide
L-Lys	L-Lysine
DL-Met	DL-Methionine
*m*RNA	Messenger ribonucleic acid
NAD	Nicotinamide-adenine dinucleotide
NADH	Nicotinamide-adenine dinucleotide (reduced form)
NADP	Nicotinamide-adenine dinucleotide phosphate
NADPH	Nicotinamide-adenine dinucleotide phosphate (reduced form)
NMN	Nicotinamide mononucleotide
OAA	Oxalacetic acid
ONPG	*o*-Nitrophenyl galactoside
PaA	Pantothenic acid
P-cellulose	Phosphoryl cellulose
Pen G	Penicillin G
PAB-cellulose	*p*-Aminobenzoyl cellulose
PLP	Pyridoxal phosphate
PMP	Pyridoxamine phosphate
PNP	Pyridoxine phosphate
PNPA	*p*-Nitrophenyl acetic acid
PNPP	*p*-Nitrophenyl phosphate
PNPS	*p*-Nitrophenyl sulfate
Poly A	Polyadenylic acid
Poly C	Polycytidylic acid
Poly U	Polyuridylic acid
Pyr	Pyruvic acid
RNA	Ribonucleic acid
SDS	Sodium dodecyl sulfate
DL-Ser	DL-Serine
TAME	α-*N*-Tosyl-L-arginine methyl ester

TEAE-cellulose	Triethylaminoethyl cellulose
TEMED	N,N,N',N'-Tetramethyl ethylenediamine
tRNA	Transfer ribonucleic acid
L-Trp	L-Tryptophan
L-Tyr	L-Tyrosine
U	Uridine
UDP	Uridine 5'-diphosphate
2',3'-cUp	Uridine 2',3'-cyclic phosphate

Subject index

Enzyme index

β-fructofuranosidase, 17, 64, 118, 128
fructosediphosphatase, 24, 39
fumarase, 50, 75, 76, 122, 134, 139–41, 159, 191, 244

G

β-galactosidase, 12, 18, 50, 76, 112, 124, 128, 137, 230
glucoamylase, 14, 15, 18, 23, 25, 26, 34, 35, 39, 40, 44, 50, 51, 60, 62, 64, 68, 69, 109, 110, 112, 124, 126, 129, 130, 137, 164, 178, 208
glucose isomerase, 12, 18, 39, 51, 64, 74–6, 80, 138, 164, 195
— oxidase, 12, 14, 17, 23, 25, 34, 40, 44, 47, 50, 62–4, 67, 124, 137, 208, 210, 211, 213, 214, 233, 253
—-6-phosphate dehydrogenase, 25, 133, 200, 220
— — isomerase, 14, 51, 199
β-glucosidase, 26, 213
β-D-—, 25, 26, 27
glutamate dehydrogenase, 16, 25, 47, 63, 197, 213
L-— —, 17, 24, 35
—-oxaloacetate transaminase, 250
—-pyruvate transaminase, 63, 197
glutaminase, 213
glyceraldehyde phosphate dehydrogenase, 24, 39, 124, 135
—-3-— —, 202

H, I

hepatic flavoprotein oxidase, 26, 166
— microsomal flavoprotein oxidase, 39, 64
hexokinase, 25, 50, 54, 62, 63, 199, 200, 220
L-histidine ammonia-lyase, 75, 119, 122, 141, 143, 193
histidyl-tRNA synthetase, 252
hydrogenase, 18, 27, 39
Δ'-—, 164
11β-hydroxylase, 166

inorganic pyrophosphatase, 14, 198
invertase, 11, 12, 14–6, 18, 23, 28, 39, 40, 44, 50, 51, 54, 63, 64, 66, 67, 69, 70, 74, 76, 80, 112, 121, 128, 133, 134, 137–9, 164, 181, 208, 213
isocitrate dehydrogenase, 25, 131
isoleucyl-tRNA synthetase, 252

K, L

kallikrein, 24

lactase, 12, 18, 25, 27, 28, 39, 40, 50, 51, 54, 62–5, 67, 69, 71, 208, 214, 220, 230
lactate dehydrogenase, 14, 17, 23–5, 35, 39, 40, 44, 50, 63, 64, 124, 129, 130, 131, 133, 135, 136, 208, 213, 218
leucine aminopeptidase, 12, 18, 25–7, 30, 129, 133, 164, 176, 215
lipase, 15, 17, 24, 28, 54, 63, 80, 122
lipoamide dehydrogenase, 12, 13
lipoxygenase, 25
lysozyme, 12, 16, 17, 27, 47, 63

M, N, O

malate aminotransferase, 208
— dehydrogenase, 25, 39, 50, 54, 63, 64, 208, 218
mannosidase, 26
L-menthol ester hydrolase, 75
multi-enzyme, 74–6
mutarotase, 213
myosin ATPase, 16

NAD kinase, 75
— pyrophosphorylase, 12, 112, 124, 198
naringinase, 12, 24, 135, 136, 234
neutral protease, 18, 26, 40, 51
nitrate reductase, 80, 208

L-ornithine transcarbamylase, 206
orsellinic acid decarboxylase, 75
D-oxynitrilase, 14, 164, 195

P

papain, 12, 16–8, 20, 23, 24, 27, 34, 35, 40, 44, 47, 50, 62, 63, 69, 70, 109, 111, 136, 138, 142, 164, 185, 215, 233, 234, 253
pectin esterase, 24
penicillin acylase, 74–6, 182
— amidase, 35, 62, 70, 119, 122, 134, 138, 141, 164, 182
penicillinase, 51, 213
pepsin, 2, 14, 17, 21, 27, 35, 39, 40, 42, 69, 70, 138
peroxidase, 25, 27, 47, 62, 63, 65, 213, 214, 234, 254
phenylalanine decarboxylase, 39
phosphodiesterase, 12, 35, 216
phosphofructokinase, 50, 199
phospholipase, 12
phosphomonoesterase, 35, 266
phosphorylase, 26, 35, 39, 137, 139, 164, 167
phosphotransacetylase, 203